SpringerBriefs in Applied Sciences and Technology

Thermal Engineering and Applied Science

Series editor

Francis A. Kulacki, Minneapolis, MN, USA

More information about this series at http://www.springer.com/series/8884

David Goluskin

Internally Heated Convection and Rayleigh-Bénard Convection

 Springer

David Goluskin
Mathematics Department and
    Center for the Study of Complex Systems
University of Michigan
Ann Arbor, MI, USA

ISSN 2191-530X   ISSN 2191-5318   (electronic)
SpringerBriefs in Applied Sciences and Technology
ISBN 978-3-319-23939-2   ISBN 978-3-319-23941-5   (eBook)
DOI 10.1007/978-3-319-23941-5

Library of Congress Control Number: 2015950838

Springer Cham Heidelberg New York Dordrecht London

Printed on acid-free paper

Springer International Publishing AG Switzerland is part of Springer Science+Business Media (www.
springer.com)

Preface

The purpose of this SpringerBrief is to review heat transfer in layers of convective fluid. Six different configurations are considered—three that are versions of Rayleigh–Bénard (RB) convection, which is driven by differential heating at the boundaries, and three that are driven by uniform internal heating. The essential features of all six models are derived mathematically. The experimental literature is reviewed in depth for the models of internally heated (IH) convection, which are much less studied than their RB counterparts. Experiments on RB convection are treated in less depth, as they have been thoroughly reviewed elsewhere.

Along with placing the various convective models within a conceptual framework that brings out their similarities, we give some minor results not published elsewhere. For instance, a few of the linear instability and energy stability thresholds given in Tables 2.2 and 2.3 either have not been reported before or have been reported with less precision. One of the bounds proven in Sect. 2.3 is also new, as are the visualizations of simulations included as Fig. 1.2.

Chapter 1 provides background and then defines the six configurations under study, the governing equations of our models, and their basic features. Chapter 2 presents results that can be derived mathematically from the governing equations: linear and nonlinear stability thresholds of static states, along with proven bounds on mean temperatures and heat fluxes. For the IH cases only, Chap. 3 gives a quantitative survey of heat transport in both laboratory experiments and numerical simulations, followed by suggestions for future work.

The author is grateful to Charles R. Doering, Erwin P. van der Poel, Jared P. Whitehead, and Francis A. Kulacki for their many helpful comments on the manuscript. Thanks also to Martin Wörner for providing his original data and to Francis A. Kulacki for providing not only his original data but also many hard-to-find references. More general thanks are due to Edward A. Spiegel, who taught the author much of what he knows about convection and many other topics.

Ann Arbor, MI, USA David Goluskin

Contents

Chapter 1
A Family of Convective Models

Abstract Six canonical models of convection are described: three configurations of internally heated (IH) convection driven by constant and uniform volumetric heating, and three configurations of Rayleigh–Bénard (RB) convection driven by the boundary conditions. The IH models are distinguished by differing pairs of thermal boundary conditions: top and bottom boundaries of equal temperature, an insulating bottom with heat flux fixed at the top, and an insulating bottom with temperature fixed at the top. The RB models too are distinguished by whether temperatures or heat fluxes are fixed at the top and bottom boundaries. Integral quantities important to heat transport are examined, including the mean fluid temperature, the mean temperature difference between the boundaries, and the mean convective heat flux. Integral relations and bounds are presented, and further bounds are conjectured for the IH cases. Similarities and differences between the six configurations are emphasized.

This SpringerBrief is concerned entirely with convection—fluid motion driven by differential body forces. We focus on several simple configurations that lend themselves to theoretical and experimental study. Convection arises in many contexts and figures prominently in astrophysics, geophysics, and certain engineering applications. Astrophysical occurrences include stellar interiors [15, 17, 29, 45] and planetary atmospheres and interiors [22, 26, 28, 41], while terrestrial occurrences include the Earth's outer core [4, 6, 7, 14], mantle [39], oceans [33, 46], and atmosphere [13]. Many of these systems have internal sources or sinks of buoyancy, including the Earth's mantle, the cores of large main-sequence stars, radiating atmospheres, and nearly any engineered system where chemical or nuclear reactions take place in a fluid environment. Among such engineering applications, particular attention has been paid to nuclear accident scenarios in which exothermic nuclear reactions drive convection in molten material [3, 21, 34].

We speak here in terms of thermal convection, where the body forces are gravitational and depend on the fluid's density, which, in turn, depends on its temperature. Other types of convection are not discussed but are often governed by similar dynamics. In compositional convection, for instance, chemical concentration takes the place of temperature. In electroconvection (e.g., [5, 47]), electric charge takes

© Springer International Publishing Switzerland 2016   1
D. Goluskin, *Internally Heated Convection and Rayleigh-Bénard Convection*,
SpringerBriefs in Applied Sciences and Technology,
DOI 10.1007/978-3-319-23941-5_1

the place of temperature, and electrical potential takes the place of gravitational potential.

Convection can be indefinitely sustained in each configuration studied here, and we focus on the time-averaged properties of sustained convection, especially heat transport. Transient phenomena are not addressed. The minimum requirement for ordinary fluid to convect is that warmer (less dense) fluid lie below cooler (more dense) fluid, and that this adverse temperature gradient be sufficiently destabilizing to overcome the viscous forces that damp fluid motion. For the convection to be indefinitely sustained against viscous dissipation, there is an additional requirement: an inexhaustible source of energy that drives the system away from equilibrium by endlessly adding heat somewhere other than the top of the domain or removing heat somewhere other than the bottom. This can be accomplished through the thermal boundary conditions, as when a pot of water is boiled on a stove, or it can be accomplished through internal heat sources or sinks, as when radioactive decay heats the Earth's mantle.

The present chapter lays out the basic features of six convective configurations— three that are driven solely by the boundary conditions, and three that are driven by internal heating. The configurations are defined in Sect. 1.1, and the Boussinesq equations that are used to model them are given in Sect. 1.2, followed in Sect. 1.3 by our chosen nondimensionalization. The basic commonalities and differences of the six configurations are then summarized: Sect. 1.4 discusses static states, Sect. 1.5 gives a qualitative look ahead to the experimental findings that are surveyed in Chap. 3, and Sect. 1.6 introduces the integral quantities and relations that govern heat transport.

1.1   Six Configurations

The six configurations we study here share the same basic geometry: a horizontal fluid layer of height d. In its horizontal dimensions, the layer can be modeled as infinite, periodic, or bounded, though only the last case is realizable in physical experiments. In a theoretical investigation, we can allow the convection to have three-dimensional (3D) freedom, or we can make it two-dimensional (2D) by imposing uniformity in one of the horizontal dimensions.

The six configurations differ only in their top and bottom thermal boundary conditions and in the presence or absence of volumetric heating. Three of the cases are versions of Rayleigh–Bénard (RB) convection, wherein the flow is driven solely by thermal boundary conditions that cause heat to enter across the bottom boundary and exit across the top one. The other three cases are instances of internally heated (IH) convection. Here, heat is added only by a constant, uniform source, and at least some of it exits across the top boundary. The only thermal boundary conditions we employ are fixed-temperature, fixed-flux, or perfectly insulating. Fixed temperatures model perfectly conductive boundaries, while fixed heat fluxes model boundaries that conduct heat poorly [43].

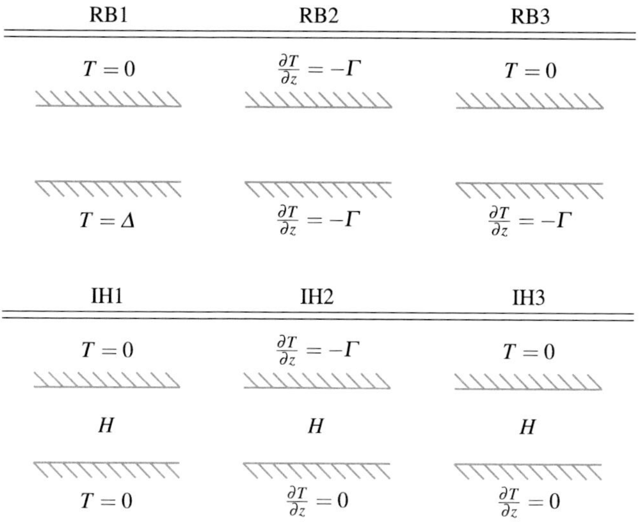

RB1	RB2	RB3
$T = 0$	$\frac{\partial T}{\partial z} = -\Gamma$	$T = 0$
$T = \Delta$	$\frac{\partial T}{\partial z} = -\Gamma$	$\frac{\partial T}{\partial z} = -\Gamma$

IH1	IH2	IH3
$T = 0$	$\frac{\partial T}{\partial z} = -\Gamma$	$T = 0$
H	H	H
$T = 0$	$\frac{\partial T}{\partial z} = 0$	$\frac{\partial T}{\partial z} = 0$

Fig. 1.1 Schematics of the six configurations studied in the present work. They are distinguished by their thermal boundary conditions and by the presence or absence of a constant and uniform internal heat source (H). Gravity acts vertically downward. All quantities are dimensional; nondimensionalized versions of these schematics appear in Tables 1.1 and 1.2. In the IH2 configuration, H and Γ are related such that heat loss balances internal heat production (see text)

The thermal boundary conditions of our three RB configurations are shown in the top row of Fig. 1.1. In the case we call RB1, which is the most-studied RB model, the boundary temperatures are fixed, with a temperature drop of Δ from the bottom boundary to the top one. (The temperature at the top boundary can be fixed at zero for convenience because, under the governing equations of our models, only temperature *differences* affect the dynamics.) In RB2, the heat flux across each boundary is fixed by setting the temperature gradients there to the same value, $-\Gamma$. The RB3 case mixes the two previous cases, having a fixed-temperature condition on the top boundary and a fixed-flux condition on the bottom one. The configuration where these two boundary conditions are swapped need not be considered since it is related to RB3 by symmetry, at least with the governing equations of our models.

The thermal boundary conditions of our three IH configurations are shown in the bottom row of Fig. 1.1. Temperature is produced volumetrically at rate H in all three cases, corresponding to heat production at rate H/c_p, where c_p is the fluid's heat capacity. In IH1, the top and bottom temperatures are fixed at the same value, and the internally produced heat escapes across both boundaries. In IH2 and IH3, the bottom boundary is perfectly insulating, meaning the vertical temperature gradient must vanish there. The top boundary, across which all the internally produced heat escapes, has a fixed flux in IH2 and a fixed temperature in IH3. In IH2, the boundary flux must match the rate of internal heat production in order for convection to be statistically steady, so Γ is determined by H, as described in Sect. 1.3. Configurations where the boundary conditions in IH2 or

IH3 are reversed, making the tops insulating, do not need to be considered; fluid in such configurations would remain static, as follows from the stability results of Sect. 2.2. Notice that the temperature is subject to Dirichlet conditions in RB1 and IH1, Neumann conditions in RB2 and IH2, and one condition of each type in RB3 and IH3.

For the velocity, we impose either no-slip or free-slip boundary conditions. No-slip conditions apply to most laboratory experiments and many engineering applications, while free-slip conditions are more appropriate in modeling certain astrophysical, geophysical, and plasma physical systems. Although we address both possibilities, some results are available only for no-slip boundaries. If side boundaries exist, we assume they are perfect thermal insulators.

The six models of Fig. 1.1 can be extended in various ways. For instance, internal heating can be added to the RB configurations, creating hybrid models driven both by the boundary conditions and by internal heating. The thermal boundary conditions can also be made more complicated, perhaps to model thermal radiation or moderate conductivity. However, such models require at least one more control parameter than those of Fig. 1.1. When modeled by the Boussinesq equations, which are introduced in the next section, each of the six configurations we study is governed by *only two* dimensionless control parameters, aside from any parameters used to describe the geometry. In a sense described at the end of Sect. 1.3, they are the only models of convective layers for which this is true, hence they are a natural starting point.

In many sections of this SpringerBrief results are presented for all six configurations in Fig. 1.1, letting us highlight their similarities and differences. Section 2.3 and Chap. 3, however, focus mainly on IH convection. This work is not meant to be a comprehensive review of RB convection, which has been reviewed several times in recent decades [1, 11, 18, 32, 40]. Interest in RB convection goes well beyond heat transport, as the system has become a canonical model of nonlinear science, having provided early examples of instabilities, bifurcations, pattern formation, and chaos in spatially extended systems. IH convection, which has been the subject of numerous works but is still much less studied than its RB counterpart, was last reviewed in the 1980s [10, 31]. Bringing this topic up to date requires not only that we review the more recent studies of IH convection, which are relatively few in number, but also that we reinterpret older studies in light of our contemporary understanding of RB convection.

Many of the laboratory experiments reviewed in Chap. 3 are captured well by one of our three IH models, but applications are rife with further complications, including compressibility, temperature-dependent material properties, complicated geometries, chemical and nuclear reactions, rotation, magnetism, and other thermal boundary conditions. Nonetheless, characterizing heat transport in the relatively simple models we consider is already a formidable challenge, and the task is far from complete.

1.2 Boussinesq Equations

A mathematical model of thermal convection must include equations governing the velocity and temperature fields, along with a constitutive relation between temperature and density. The compressible Navier–Stokes equations [50] describe the velocity field accurately in a wide range of physical applications, but they are challenging equations to study analytically or integrate numerically, and they can be avoided when the density field does not deviate too strongly from hydrostatic equilibrium. Pressure-driven flows with weak density variations can simply be approximated as having constant densities, yielding the incompressible Navier–Stokes equations. In buoyancy-driven flows, however, density variations cannot be totally ignored because they create the buoyancy gradients needed to drive motion. The typical compromise is to employ the Boussinesq approximation, as we do here.

The Boussinesq approximation, which was first invoked by Oberbeck [35] and is also called the Oberbeck–Boussinesq approximation, involves two main assumptions. First, the fluid's density, ρ, is assumed to vary linearly with temperature, T, about some hydrostatic reference state denoted by ρ_* and T_*. That is,

$$\rho(T) = \rho_*\left[1 - \alpha(T - T_*)\right], \tag{1.1}$$

where α is the linear coefficient of thermal expansion. Second, the density variations are assumed to be sufficiently weak that they can be ignored everywhere except in the buoyancy force. The fluid is sometimes called incompressible since the velocity field is divergence-free, although compressibility does manifest in the buoyancy variations. Numerous justifications have been put forth for replacing the fully compressible Navier–Stokes equations with the simpler Boussinesq equations, typically invoking some combination of asymptotic expansions and ad hoc assumptions. See Spiegel and Veronis [44] for one physical justification and Rajagopal et al. [37] for a discussion of various other justifications. The precise assumptions invoked vary, but in all versions there is a sense in which gradients of the fluid's properties should not be too steep. If the Boussinesq approximation is used in modeling a physical system, the assumptions under which the approximation holds should be checked if possible, either by physical measurement or by numerical simulation of compressible equations.

With constant gravitational acceleration g acting in the $-\hat{\mathbf{z}}$ direction, applying the Boussinesq approximation to the compressible Navier–Stokes equations yields the Boussinesq equations [8, 38],

$$\nabla \cdot \mathbf{u} = 0 \tag{1.2}$$

$$\partial_t \mathbf{u} + \mathbf{u} \cdot \nabla \mathbf{u} = -\frac{1}{\rho_*}\nabla p + \nu\nabla^2\mathbf{u} + g\alpha T\hat{\mathbf{z}} \tag{1.3}$$

$$\partial_t T + \mathbf{u} \cdot \nabla T = \kappa\nabla^2 T + H, \tag{1.4}$$

where $\mathbf{u} = (u, v, w)$ is the fluid's velocity vector, p its pressure, ν its kinematic viscosity, and κ its thermal diffusivity. The temperature source term, H, is absent from RB convection but drives the convection in our IH models. The pressure term in (1.3) has absorbed hydrostatic terms of the buoyancy force coming from (1.1).

1.3   Nondimensionalization

To nondimensionalize the Boussinesq equations, we scale distance by the layer height, d, time by the characteristic timescale of thermal diffusion, d^2/κ, and pressure by $\rho_* d^2/\kappa$. We scale temperature by a dimensional quantity, Δ, that is defined differently in various configurations. In the RB1 case, Δ is the prescribed temperature difference between the boundaries. In the other cases,

$$\Delta := \begin{cases} d\Gamma & \text{RB2, RB3, IH2} \\ \frac{d^2 H}{\kappa} & \text{IH1, IH2, IH3.} \end{cases} \tag{1.5}$$

Nondimensionalized by these Δ, the temperature difference between the boundaries is unity in RB1; the fixed temperature fluxes are unity in RB2, RB3, and IH2; and the volumetric heating rate is unity in the IH cases. Both definitions in (1.5) apply to IH2 because we add the consistency condition $\Gamma = dH/\kappa$ in that case to ensure that heat production balances heat loss.

The Boussinesq equations (1.2)–(1.4) in dimensionless form are

$$\nabla \cdot \mathbf{u} = 0 \tag{1.6}$$

$$\partial_t \mathbf{u} + \mathbf{u} \cdot \nabla \mathbf{u} = -\nabla p + Pr\nabla^2 \mathbf{u} + PrRT\hat{\mathbf{z}} \tag{1.7}$$

$$\partial_t T + \mathbf{u} \cdot \nabla T = \nabla^2 T + Q, \tag{1.8}$$

where the symbols $\mathbf{u}$, T, p, $\mathbf{x}$, and t henceforth represent dimensionless quantities. The vertical extent is $0 \leq z \leq 1$, and the temperature source term is

$$Q = \begin{cases} 0 & \text{RB} \\ 1 & \text{IH.} \end{cases} \tag{1.9}$$

The dimensionless control parameters are the Rayleigh number, R, and Prandtl number, Pr, defined by

$$R := \frac{g\alpha d^3 \Delta}{\kappa \nu} \tag{1.10}$$

$$Pr := \frac{\nu}{\kappa}. \tag{1.11}$$

The definition of R differs between cases when the definition (1.5) of Δ differs.

The Rayleigh number may be thought of as the ratio of inertial forces to viscous forces. When it is large, the fluid is strongly driven by differential buoyancy forces. We regard R as the primary control parameter since raising it typically makes the flow more complex. For R to indeed be a control parameter, we needed to define the dimensional temperature scale, Δ, in terms of quantities that are known a priori: the boundary conditions or heating rate. However, it is sometimes useful to define a different Rayleigh number using a temperature scale that is determined dynamically by the flow. This sort of Rayleigh number cannot serve as a control parameter but can be a useful diagnostic quantity. Thus, we will sometimes distinguish between the *control* Rayleigh number, R, and *diagnostic* Rayleigh numbers, Ra or $\widetilde{Ra}$, defined in Sect. 1.6.5.

The Prandtl number is the rate at which the fluid diffuses momentum, relative to the rate at which it diffuses heat. Unlike the Rayleigh number, it is a material property of the fluid and does not depend on the geometry or boundary conditions. The Prandtl number is large in fluids that damp motion strongly and conduct heat poorly. In the Earth's mantle, for instance, Pr is effectively infinite. The Prandtl number is small in fluids that damp motion weakly and conduct heat well, such as liquid metals and stellar plasmas. Air and water are intermediate examples, having Prandtl number close to 0.7 and 7, respectively, under atmospheric conditions.

In all six configurations of Fig. 1.1, the dynamics depend on only two control parameters, Pr and R, aside from any parameters describing the geometry, such as aspect ratios. This is the minimum number of parameters we can hope for in the study of convection, except in those special cases where Pr can be eliminated because it is effectively infinite. Additional parameters would be needed if the thermal boundary conditions were more complicated [24, 43], the internal heating law were more complicated [16, 48], or the boundary conditions and internal heating each introduced their own temperature scales [2, 9, 23, 27, 30, 42, 52]. We restrict ourselves to models that require only Pr and R since every additional parameter makes it much harder to understand parameter space. In fact, among the ways that uniform heating, fixed-temperature boundaries, and fixed-flux boundaries can be combined, our six configurations (and their symmetry-related siblings) seem to be the only ones governed by so few parameters.

1.4 Static States

Dimensionless schematics of the RB and IH configurations are shown in the first rows of Tables 1.1 and 1.2, respectively. The basic features of the six cases are summarized in the subsequent rows of both tables and are further laid out in the remainder of this chapter. The simplest solutions to the governing equations are static, with heat transported only by conduction. These are the unique asymptotic states when R is sufficiently small (cf. Sect. 2.2), and they solve the Poisson or Laplace equation $\nabla^2 T + Q = 0$. Since we assume that side boundaries are nonexistent or perfectly insulating, the static temperature fields, T_{st}, vary only in z:

Table 1.1 Summary of the properties of RB convection discussed in the present chapter

	RB1	RB2	RB3
Configuration	$T=0$ \\\\\\\\\\\ ////////// $T=1$	$\frac{\partial T}{\partial z}=-1$ \\\\\\\\\\\ ////////// $\frac{\partial T}{\partial z}=-1$	$T=0$ \\\\\\\\\\\ ////////// $\frac{\partial T}{\partial z}=-1$
Static temperature profile			
Turbulent temperature profile			
Heat balance	$\frac{d}{dz}\overline{T}\big|_{z=1} = \frac{d}{dz}\overline{T}\big|_{z=0}$		
$\overline{J}(z)$	$1+\langle wT\rangle$	1	
$\langle J\rangle$	$1+\langle wT\rangle$	1	
Uniform $\langle wT\rangle$ bounds	$0\le\langle wT\rangle<\infty$	$0\le\langle wT\rangle<1$	
Uniform $\delta\langle T\rangle$ bounds	$0<\delta\langle T\rangle<1$	$-\frac{1}{\sqrt{3}}\le\delta\langle T\rangle\le\frac{1}{\sqrt{3}}$	$0<\delta\langle T\rangle\le\frac{1}{\sqrt{3}}$
Empirical relation	$\delta\langle T\rangle\sim\frac{1}{2}$	$\delta\langle T\rangle\sim\frac{1}{2}\left(1-\langle wT\rangle\right)$	
N	$1+\langle wT\rangle$	$\dfrac{1}{1-\langle wT\rangle}$	

All quantities are dimensionless, and the vertical extent is $0\le z\le 1$. Notation is defined throughout the chapter. Briefly, $\overline{*}$ denotes an average over horizontal directions and time, $\langle *\rangle$ denotes an average over volume and time, $\delta\langle T\rangle$ is the mean temperature of the fluid relative to that of the top boundary, and $J=-\partial_z T+wT$ is the sum of the conductive and convective vertical heat fluxes

Table 1.2 Summary of the properties of IH convection discussed in the present chapter

	IH1	IH2	IH3
Configuration	$T = 0$ $Q = 1$ $T = 0$	$\frac{\partial T}{\partial z} = -1$ $Q = 1$ $\frac{\partial T}{\partial z} = 0$	$T = 0$ $Q = 1$ $\frac{\partial T}{\partial z} = 0$
Static temperature profile	(profile; $\frac{1}{8}$)	(profile; $\frac{1}{2}$)	(profile; $\frac{1}{2}$)
Turbulent temperature profile	(profile)	(profile)	(profile)
Heat balance	$-\frac{d}{dz}\overline{T}\big|_{z=1} + \frac{d}{dz}\overline{T}\big|_{z=0} = 1$	$-\frac{d}{dz}\overline{T}\big|_{z=1} = 1$	
$\overline{J}(z)$	$\langle wT\rangle + \left(z - \frac{1}{2}\right)$	z	
$\langle J\rangle$	$\langle wT\rangle$	$\frac{1}{2}$	
Uniform $\langle wT\rangle$ bounds	$0 \le \langle wT\rangle < \frac{1}{2}$	$0 \le \langle wT\rangle < \frac{1}{2} + \frac{1}{\sqrt{3}}$	$0 \le \langle wT\rangle < \frac{1}{2}$
Uniform $\delta\langle T\rangle$ bounds	$0 < \delta\langle T\rangle \le \frac{1}{12}$	$0 < \delta\langle T\rangle \le \frac{1}{3}$	
Empirical relation	$\delta\langle T\rangle \sim \overline{T}_{max}$	$\delta\langle T\rangle \sim \frac{1}{2} - \langle wT\rangle$	
N	$\dfrac{1}{8\overline{T}_{max}}$	$\dfrac{1}{1 - 2\langle wT\rangle}$	
$\widetilde{N}$	$\dfrac{1}{12\,\delta\langle T\rangle}$	$\dfrac{1}{3\,\delta\langle T\rangle}$	

All quantities are dimensionless, and the vertical extent is $0 \le z \le 1$. Notation is defined throughout the chapter. Briefly, $\overline{*}$ denotes an average over horizontal directions and time, $\langle *\rangle$ denotes an average over volume and time, $\delta\langle T\rangle$ is the mean temperature of the fluid relative to that of the top boundary, $J = -\partial_z T + wT$ is the sum of conductive and convective vertical heat fluxes, and $\overline{T}_{max}$ is the maximum value of $\overline{T}(z)$

$$T_{st}(z) = \begin{cases} 1-z & \text{RB1, RB2, RB3} \\ \frac{1}{2}z(1-z) & \text{IH1} \\ \frac{1}{2}(1-z^2) & \text{IH2, IH3.} \end{cases} \qquad (1.12)$$

These purely conductive profiles are depicted in the second rows of Tables 1.1 and 1.2. They are parabolic with internal heating and linear without it. The static profiles in RB2 and IH2 are determined only up to additive constants, but these constants do not affect the dynamics, so we have fixed them for convenience.

In each configuration, our dimensional temperature scale, Δ, is characteristic of the static state. This is why the dimensionless T_{st} have no dependence on R. When we define *diagnostic* Rayleigh number in Sect. 1.6.5, we will do so by replacing Δ with temperature scales characteristic of the convective states, rather than the static ones.

1.5   Temperature Fields in Strong Convection

Whereas the fluid remains static when R is sufficiently small, it convects strongly when R is large. Convection typically strengthens monotonically as R is raised, though this is not universally true and can depend on how strength is quantified. (The non-monotonicity of convective transport in [20] provides a counterexample.) Figure 1.2 shows instantaneous temperature fields from 2D simulations at large R. The RB1 field in Fig. 1.2a is representative of all three RB cases: hot plumes rise from the bottom boundary, cold plumes descend from the top one, and both types of plumes contribute to upward heat transport. In the IH1 field of Fig. 1.2b, cold plumes descend from the top, but the bottom boundary layer is cold and *stably* stratified. This bottom layer emits no buoyant plumes, so any mixing between it and interior must be driven by shear, rather than buoyancy. The IH3 field in Fig. 1.2c is representative of both IH2 and IH3: cold plumes descend from the top boundary, and there is no thermal boundary layer at the bottom.

The turbulent convection that occurs at large R creates mean vertical temperature profiles very different from the static ones. Rough schematics of such profiles are shown in the third rows of Tables 1.1 and 1.2. Although these schematics include no secondary details, they illustrate the main differences between the various configurations. In each of the six cases, mixing by strong convection tends to flatten the temperature profile in the layer's interior. (We are not aware of a counterexample in 3D, though one exists in 2D under conditions that allow very strong winds to develop [20, 51].) The roughly isothermal interiors are flanked by one or two thermal boundary layers, and these are what distinguish the various cases.

In all six of our models, temperature is unstably stratified at the top boundary. At the bottom boundary, the temperature is unstably stratified in the RB cases, stably stratified in IH1, and unstratified in IH2 and IH3. These facts are evident in the static temperature profiles (cf. Tables 1.1 and 1.2) and remain provably true of sustained convection, at least in a time-averaged sense. As convective heat transport rises, the

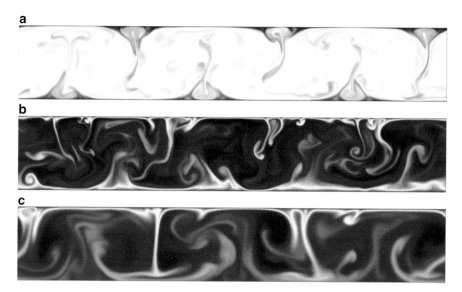

Fig. 1.2 Temperature fields from 2D simulations of (**a**) RB1, (**b**) IH1, and (**c**) IH3. Each simulation employed a horizontal period of 6, no-slip boundaries above and below, $Pr = 1$, and $R/R_L = 10^5$, where R_L is the Rayleigh number at which the static becomes linearly unstable (cf. Sect. 2.1). The coolest fluid (*blue*) has a temperature of zero in each case, and the warmest fluid (*red*) has a temperature of 1, 0.017, and 0.044, respectively

mean temperature profiles undergo various changes. In RB1, where the temperature difference between the boundaries cannot change, the boundary layers steepen as the heat flux through the domain rises. In the other two RB cases, where the mean flux through the domain cannot change, the temperature difference between the boundaries decreases. In the IH cases, the temperature of the fluid, relative to that of the top boundary, drops as the convection strengthens. The produced heat leaves only across the top boundary in IH2 and IH3. It leaves across both boundaries in IH1, but the majority leaves across the top boundary, hence the top boundary layer is steeper than the bottom one.

Although the turbulent temperature profiles are fairly easy to understand qualitatively, it is very difficult in general to anticipate their quantitative features. This would be tantamount to accomplishing our main objective of characterizing the bulk heat transport.

1.6   Mean Heat Fluxes and Integral Relations

Heat in a convecting fluid is transported by two mechanisms simultaneously: conduction and convection. Conduction refers to the diffusion of heat down the temperature gradient, while convection refers to the advection of heat by fluid motion. The temperature equation (1.8) can be written in the standard form of a conservation law as $\partial_t T + \nabla \cdot \mathbf{J} = Q$, where $\mathbf{J} := \mathbf{u}T - \nabla T$ is the total *heat current* at a point. Evidently, the conductive current is $-\nabla T$ in our nondimensionalization, and the convective current is $\mathbf{u}T$. The horizontal components of $\mathbf{J}$ vanish when averaged over the volume since our side boundaries are insulting or nonexistent. Here we focus on the heat current's vertical component, J, where

$$J := J_{\text{cond}} + J_{\text{conv}} \tag{1.13}$$

$$:= -\partial_z T + wT. \tag{1.14}$$

Much of our effort is devoted to quantifying the relative contributions to vertical heat transport made by conduction and convection—that is, by $-\partial_z T$ and wT.

In our notation, an overbar, as in $\bar{f}$, denotes an average over horizontal surfaces and infinite time. Angular brackets, as in $\langle f \rangle$, denote an average over the entire volume and infinite time. When the dimensionless domain is bounded horizontally by $0 \leq x \leq L_x$ and $0 \leq y \leq L_y$,

$$\bar{f}(z) := \lim_{\tau \to \infty} \frac{1}{\tau} \frac{1}{L_x L_y} \int_0^\tau dt \int_0^{L_y} dy \int_0^{L_x} dx f(\mathbf{x}, t), \tag{1.15}$$

$$\langle f \rangle := \lim_{\tau \to \infty} \frac{1}{\tau} \frac{1}{L_x L_y} \int_0^\tau dt \int_0^1 dz \int_0^{L_y} dy \int_0^{L_x} dx f(\mathbf{x}, t). \tag{1.16}$$

The above limits can be replaced with lim inf or lim sup to ensure their existence. For simplicity, we assume in our calculations that infinite-time averages commute with vertical averages and that horizontal averages commute with vertical derivatives, though these assumptions can often be avoided. In the above notation, the mean heat flux across a horizontal surface is

$$\bar{J}(z) = -\bar{T}'(z) + \overline{wT}(z), \tag{1.17}$$

where the prime indicates differentiation in z. The mean vertical flux across the entire layer is

$$\langle J \rangle = \delta \bar{T} + \langle wT \rangle, \tag{1.18}$$

where the mean conductive flux

$$\delta \bar{T} := \bar{T}_B - \bar{T}_T \tag{1.19}$$

is the difference between the mean bottom temperature, $\bar{T}_B$, and mean top temperature, $\bar{T}_T$. Expressions (1.17) and (1.18) are simply averages of the definition (1.14) of J; configuration-specific constraints on $\bar{J}(z)$ and $\langle J \rangle$ are given in Sect. 1.6.2.

1.6.1   Heat Balances

Conservation of heat energy is expressed in the various cases by the heat balances

$$\overline{T}'_T = \overline{T}'_B \qquad\qquad \text{RB1, RB2, RB3} \qquad\qquad (1.20)$$

$$-\overline{T}'_T + \overline{T}'_B = 1 \qquad\qquad \text{IH1} \qquad\qquad (1.21)$$

$$-\overline{T}'_T = 1 \qquad\qquad \text{IH2, IH3.} \qquad\qquad (1.22)$$

These balances, which are shown also in the fourth rows of Tables 1.1 and 1.2, are derived by averaging the temperature equation (1.8) over volume and time. Time derivatives vanish from such averages because the instantaneous volume averages of $\mathbf{u}$ and T are bounded uniformly in time (cf. Sect. 2.3). In the RB cases, the balance reflects the fact that mean heat flux into the layer at the bottom $(-\overline{T}'_B)$ must equal the mean flux out of the layer at the top $(-\overline{T}'_T)$. In IH1, the rate of internal heat production (unity) is balanced by the combined outward fluxes of heat across the top boundary $(-\overline{T}'_T)$ and the bottom one $(\overline{T}'_B)$. In IH2 and IH3, where the bottom boundary is insulating, the internal production is balanced entirely by the outward flux across the top boundary $(-\overline{T}'_T)$.

1.6.2   Constraints on Net Heat Fluxes

Little can be said a priori about the variation with height of the mean heat flux components, $-\overline{T}'(z)$ and $\overline{wT}(z)$, but we can derive constraints on their sum, $\overline{J}(z)$. Averaging the temperature equation (1.8) horizontally, vertically from 0 to z, and temporally gives

$$\overline{J}(z) := -\overline{T}'(z) + \overline{wT}(z) = \begin{cases} -\overline{T}'_B & \text{RB1} \\ 1 & \text{RB2, RB3} \\ -\overline{T}'_B + z & \text{IH1} \\ z & \text{IH2, IH3.} \end{cases} \qquad (1.23)$$

The net vertical flux is the same at every height in the RB cases and increases linearly with height in the IH cases. In the four cases where a boundary flux is fixed, $\overline{J}(z)$ is known exactly. In RB1 and IH1, $\overline{J}(z)$ is determined only up to the mean heat flux at the bottom boundary, $-\overline{T}'_B$. Below we give some alternate expressions for $\overline{J}(z)$ in these two cases, but they all involve quantities that, like $-\overline{T}_B$, are not known a priori.

The volume-averaged heat flux, $\langle J \rangle$, is also constrained. In the four cases where $\overline{J}(z)$ is known exactly, $\langle J \rangle$ is found by vertically integrating (1.23). In RB1 and IH1, the fixed-temperature boundary conditions ensure that the mean conductive fluxes are $\delta\overline{T} = 1$ and $\delta\overline{T} = 0$, respectively. The results are

$$\langle J \rangle := \delta\overline{T} + \langle wT \rangle = \begin{cases} 1 + \langle wT \rangle & \text{RB1} \\ 1 & \text{RB2, RB3} \\ \langle wT \rangle & \text{IH1} \\ \frac{1}{2} & \text{IH2, IH3.} \end{cases} \tag{1.24}$$

In all six cases, we would like to know how the control parameters affect the convective flux, $\langle wT \rangle$. In RB1 and IH1, where $\delta\overline{T}$ is fixed, this is equivalent to knowing $\langle J \rangle$. In the other four cases, where $\langle J \rangle$ is fixed, it is equivalent to knowing $\delta\overline{T}$.

In the RB1 and IH1 cases, the expressions (1.23) for $\overline{J}(z)$ can be rewritten by replacing $\overline{T}'_B$ with other integral quantities. Relations between $\overline{T}'_B$ and $\overline{T}'_T$ are provided by the heat balances of Sect. 1.6.1, while relations between $\overline{T}'_B$ and $\langle wT \rangle$ are found by equating the $\langle J \rangle$ expressions (1.24) with the vertical integrals of the $\overline{J}(z)$ expressions (1.23). The alternate expressions for $\overline{J}(z)$ found in this way are

$$\overline{J}(z) = \begin{cases} -\overline{T}'_B & = -\overline{T}'_T & = 1 + \langle wT \rangle & \text{RB1} \\ -\overline{T}'_B + z & = -\overline{T}'_T - (1-z) & = \left(z - \frac{1}{2}\right) + \langle wT \rangle & \text{IH1.} \end{cases} \tag{1.25}$$

The constraints on $\overline{J}(z)$ and $\langle J \rangle$, expressed in terms of volume integrals, are summarized in the fifth and six rows of Tables 1.1 and 1.2.

In the IH1 configuration, there is yet another useful way to interpret $\langle wT \rangle$: in terms of the *fractions* of internally produced heat that flow outward across the top and bottom boundaries. Expressions for these fractions, which we call $\mathscr{F}_T$ and $\mathscr{F}_B$, follow from relations (1.21) and (1.25),

$$\mathscr{F}_T = -\overline{T}'_T = \tfrac{1}{2} + \langle wT \rangle \tag{1.26}$$

$$\mathscr{F}_B = \overline{T}'_B = \tfrac{1}{2} - \langle wT \rangle. \tag{1.27}$$

The top and bottom fractions are both 1/2 in the static state, but convective transport breaks this symmetry.

1.6.3 $\langle wT \rangle$ and $\delta\langle T \rangle$

Essential information about heat transport is captured by the volume integrals $\langle wT \rangle$ and $\delta\langle T \rangle$, where

$$\delta\langle T \rangle := \left\langle T - \overline{T}_T \right\rangle \tag{1.28}$$

is the mean fluid temperature, relative to that of the top boundary. The above definition is needed only for RB2 and IH2, where the top temperature is not fixed. In the other configurations, $\delta\langle T\rangle \equiv \langle T\rangle$ since we have set $T_T \equiv 0$. Neither $\langle wT\rangle$ nor $\delta\langle T\rangle$ is known a priori when the fluid is flowing. Instead, the quantities must be studied by physical and computational experiments, and they can sometimes be bounded analytically. All bounds and many experimental findings that we review in the following chapters can be stated in terms of $\langle wT\rangle$ or $\delta\langle T\rangle$. In the literature, however, results on $\langle wT\rangle$ are often stated differently but equivalently in terms of other quantities, including $\delta\overline{T}$, $\overline{T}_B'$, $\overline{T}_T'$, and the Nusselt number N defined in Sect. 1.6.4 below.

1.6.3.1   Uniform Bounds

The seventh and eighth rows of Tables 1.1 and 1.2 give uniform bounds on $\langle wT\rangle$ and $\delta\langle T\rangle$—that is, bounds that are independent of R and Pr. The bounds are derived in this chapter's appendix. The lower bounds on $\langle wT\rangle$ are tight in all six configurations. The upper bounds are thought to be tight (among uniform bounds), except in the IH2 case, where we suspect a uniform upper bound of 1/2. In the IH cases, the upper bounds on $\delta\langle T\rangle$ are tight, and the lower bounds are thought to be. In the RB cases, on the other hand, it is not clear whether any of the bounds on $\delta\langle T\rangle$ are tight.

The mean convective flux, $\langle wT\rangle$, saturates its lower bound of zero in each configuration only in the static state. Physically, this is because $\langle wT\rangle$ is proportional to the work exerted by buoyancy, and when motion persists this work must be positive to balance viscous dissipation. Mathematically, the positivity of $\langle wT\rangle$ in sustained convection follows from relation (2.37) in the next chapter. The upper bounds on $\langle wT\rangle$ correspond to limits in which convective transport is infinitely stronger than conductive transport. In the RB1 case, where $\delta\overline{T} = 1$, this limit is approached when $\langle wT\rangle$ grows without bound. In the four cases where the total heat flux, $\langle J\rangle$, is fixed, this limit is approached as $\langle wT\rangle \to \langle J\rangle$ (and $\delta\overline{T} \to 0$). The IH1 case is different in that $\delta\overline{T} = 0$, so $\langle wT\rangle$ is solely responsible for the mean vertical flux. However, if one thinks of the outward transport of heat across both boundaries, rather than upward transport, then the upper limit $\langle wT\rangle \to 1/2$ indeed means that convection fully takes over from conduction. The corresponding limits of the top and bottom flux fractions (1.26)–(1.27) are $\mathscr{F}_T \to 1$ and $\mathscr{F}_B \to 0$.

The mean temperature relative to that of the top boundary, $\delta\langle T\rangle$, is bounded above and below in all six cases, but the IH bounds differ in character from the RB1 bounds. The RB1 bounds given in Table 1.1 ensure that the mean temperature of the fluid lies between those of the top and bottom boundaries. The same may be true of $\delta\langle T\rangle$ in RB2 and RB3, but the bounds derived in this chapter's appendix are too crude to show it. In RB convection, the mean fluid temperature is exactly halfway between the boundary temperatures in the static state. The same is often true when the fluid is flowing, at least with symmetric boundary conditions, but it seems no rigorous statements have been proven that reflect this observation. In the IH cases, $\delta\langle T\rangle$ saturates the upper bounds of Table 1.2 only in the static states and is strictly smaller when the fluid is flowing. Its lower bound of zero, much like the uniform upper bounds on $\langle wT\rangle$, corresponds to convection being infinitely stronger than conduction. The R-dependent bounds of Sect. 2.3 show that $\delta\langle T\rangle$ could approach zero only as $R \to \infty$.

In IH convection, where R is proportional to the dimensional heating rate, H, it might seem counterintuitive that raising R tends to decrease $\delta\langle T\rangle$. However, the *dimensional* mean temperature, $\delta\langle T\rangle \Delta$, indeed rises with H. The dimensionless quantity $\delta\langle T\rangle$ falls as convection strengthens because it has essentially been normalized by its static value.

1.6.3.2 *R*-Dependent Bounds

Since many of the bounds on $\langle wT\rangle$ and $\delta\langle T\rangle$ given in Tables 1.1 and 1.2 are tight among uniform bounds, improving them requires finding bounds that depend explicitly on R or Pr. Some R-dependent bounds have been proven for the configurations we are considering, as summarized in Table 1.3. Bounds that vary with Pr have been proven recently for RB1 [12] but not yet for other cases, although some bounds have been proven for the infinite-Pr limit that are tighter than the corresponding uniform-in-Pr results, as discussed in Sect. 2.3.

In RB convection, the R-dependent bounds that have been proven are all upper bounds on $\langle wT\rangle$. They approach the uniform upper bounds as $R \to \infty$ but are tighter at all finite R. The uniform lower bounds of $0 \le \langle wT\rangle$ cannot be improved upon, if they are to hold for all solutions, since they are saturated by the static states. These states are unstable at large R, however, and $\langle wT\rangle$ typically grows with R in experiments and simulations. There might exist better lower bounds that hold only for attracting states, rather than all solutions, but we lack the mathematical machinery to prove them as yet.

In IH convection, the R-dependent bounds that have been proven are all lower bounds on $\delta\langle T\rangle$. They approach the uniform lower bounds of zero as $R \to \infty$ but are tighter at all finite R. No R-dependent upper bounds on $\langle wT\rangle$ have been proven in IH convection, but we argue in Sect. 1.6.3.4 that such proofs should be possible. On the other hand, the uniform upper bounds on $\delta\langle T\rangle$ and lower bounds on $\langle wT\rangle$ are saturated by the static states, so any efforts to improve them run into the same obstacle as efforts to improve the lower bounds on $\langle wT\rangle$ in RB convection.

Table 1.3 Proven bounds on $\langle wT\rangle$ and $\delta\langle T\rangle$ that hold at large R. The constant c differs between cases. Details and references are given in Sect. 2.3

	R-dependent bound on $\langle wT\rangle$	R-dependent bound on $\delta\langle T\rangle$
RB1	$\langle wT\rangle \le cR^{1/2}$	None
RB2, RB3	$\langle wT\rangle \le 1 - cR^{-1/3}$	None
IH1, IH2, IH3	None	$\delta\langle T\rangle \ge cR^{-1/3}$

1.6.3.3 Empirical Approximate Relations

In addition to the exact integral relations and bounds discussed above, experiments and simulations suggest some approximate relations for $\delta\langle T\rangle$ at large R. These relations are summarized in the ninth rows of Tables 1.1 and 1.2. Most can be expressed as relations between $\delta\langle T\rangle$ and $\langle wT\rangle$, at least approximately, but the underlying assumption in IH convection differs from that in RB convection.

In the RB cases, the underlying assumption is that the top and bottom boundary layers are roughly symmetric. This implies that the mean fluid temperature is about halfway between the top and bottom temperatures, as in the schematics of turbulent temperature profiles in Table 1.1. That is,

$$\delta\langle T\rangle \sim \tfrac{1}{2}\delta\overline{T} = \begin{cases} \tfrac{1}{2} & \text{RB1} \\ \tfrac{1}{2}\left(1-\langle wT\rangle\right) & \text{RB2, RB3} \end{cases} \tag{1.29}$$

for large R. These relations are not expected to hold exactly when the top and bottom boundary conditions differ, but they could nonetheless be approached as $R \to \infty$. It might be possible to prove precise versions of the above statements, such as upper and lower bounds on $\delta\langle T\rangle$ that converge to $\delta\overline{T}$ as $R \to \infty$, but we are not aware of any such results.

In the IH cases, the underlying assumption is that mean temperature profile, $\overline{T}(z)$, at large R is roughly isothermal outside of thin thermal boundary layers, as in the schematics of turbulent temperature profiles in Table 1.2. Experimental support of this assumption is presented in Chap. 3. Approximate isothermally in the IH cases implies that

$$\delta\langle T\rangle \sim \begin{cases} \overline{T}_{\max} & \text{IH1} \\ \delta\overline{T} = \tfrac{1}{2} - \langle wT\rangle & \text{IH2, IH3} \end{cases} \tag{1.30}$$

for large R, where $\overline{T}_{\max}$ is the maximum of the mean temperature profile $\overline{T}(z)$. In IH1, the assumption of an isothermal interior does not give a relation between $\delta\langle T\rangle$ and $\langle wT\rangle$, nor is any simple relation suggested by experiments.

1.6.3.4 Conjectured Upper Bounds on $\langle wT\rangle$ in IH Convection

In IH2 and IH3, the empirical observation that $\delta\langle T\rangle \sim \delta\overline{T}$ at large R suggests that the two quantities might obey similar bounds. Since bounds of the form $\delta\langle T\rangle \geq cR^{-1/3}$ have been proven, it seems likely that bounds of the form $\delta\overline{T} \geq cR^{-1/3}$ could be proven also. The latter can be alternately stated as upper bounds on $\langle wT\rangle$:

Conjecture 1. In the IH2 and IH3 configurations, there exists a constant $c > 0$ such that for all Pr and sufficiently large R,

$$\langle wT\rangle \leq \tfrac{1}{2} - cR^{-1/3}.$$

In IH1, experiments suggest that the growth of $\langle wT \rangle$ with R is similarly bounded above (cf. Chap. 3). However, since no empirical relation between $\delta \langle T \rangle$ and $\langle wT \rangle$ is apparent, the proven lower bound on $\delta \langle T \rangle$ does not suggest a form for an upper bound on $\langle wT \rangle$. We speculate that the upper bound should approach the uniform bound of $1/2$ algebraically as $R \to \infty$, but we cannot anticipate the algebraic power:

Conjecture 2. In the IH1 configuration, there exist constants $c > 0$ and $\alpha > 0$ such that for all Pr and sufficiently large R,

$$\langle wT \rangle \leq \tfrac{1}{2} - cR^{-\alpha}.$$

1.6.4 Nusselt Numbers

The relative strengths of convective and conductive heat transport are often quantified using dimensionless Nusselt numbers. In RB convection, the typically used definitions of Nusselt numbers can all be expressed in terms of $\langle wT \rangle$. In IH convection, dimensionless quantities resembling the RB Nusselt numbers can be defined in various ways by invoking $\langle wT \rangle$, $\delta \langle T \rangle$, or $\overline{T}_{\max}$. Here we consider two ways of defining Nusselt-number-like quantities, N and $\tilde{N}$. The quantity N is determined by $\overline{T}_{\max}$ in IH1 and by $\langle wT \rangle$ in the other five cases, while the quantity $\tilde{N}$ is determined in the IH cases by $\delta \langle T \rangle$.

1.6.4.1 The Nusselt Number N

The definition of N that we choose is one that has helped reveal parallels between various RB configurations when used in concert with the quantity Ra defined in the next subsection [25, 36, 53, 54]. In every case other than IH1, our definition of N can be expressed as the ratio of mean total heat flux to mean conductive heat flux, where both quantities are averages over volume and time in the developed flow,

$$N = \frac{\langle J \rangle}{\langle J_{\text{cond}} \rangle} = \frac{\delta \overline{T} + \langle wT \rangle}{\delta \overline{T}} \qquad \text{RB1, RB2, RB3, IH2, IH3.} \qquad (1.31)$$

The above definition would fail for IH1 because its denominator would be zero. Applying the various constraints on $\delta \overline{T}$ and $\langle wT \rangle$ (cf. Tables 1.1 and 1.2) to expression (1.31) and adding an ad hoc definition for the IH1 case, we obtain

$$N := \begin{cases} 1 + \langle wT \rangle & \text{RB1} \\[4pt] \dfrac{1}{\delta \overline{T}} = \dfrac{1}{1 - \langle wT \rangle} & \text{RB2, RB3} \\[8pt] \dfrac{1}{8 \overline{T}_{\max}} & \text{IH1} \\[6pt] \dfrac{1}{2 \delta \overline{\overline{T}}} = \dfrac{1}{1 - 2\langle wT \rangle} & \text{IH2, IH3.} \end{cases} \qquad (1.32)$$

The rationale for our definition of N in the IH1 case, the only case where heat flows outward across both boundaries, is that we are considering outward heat fluxes instead of upward fluxes. The mean total outward flux is unity since it must balance heat production. To determine the mean outward conduction, we imagine dividing the layer at the height z^* where the temperature profile $\overline{T}(z)$ assumes its maximum value of $\overline{T}_{max}$. The upward conduction above z^* is proportional to $\overline{T}_{max}$, as is the downward conduction below z^*. Thus, the ratio of total outward transport to conductive outward transport is inversely proportional to $\overline{T}_{max}$. The $1/8$ factor makes N unity in the static state. Although the analogy between N in IH1 and in the other five cases is not perfect, the experiments reviewed in Chap. 3 reveal similarities in N between all cases. In the IH3 case, the quantity we call N has been considered under various names, perhaps first by Thirlby [49]. In the IH1 case, our definition has apparently not been used, but many authors have considered $\overline{T}_{max}$, as well as the so-called top and bottom Nusselt numbers discussed in Sect. 3.2.2.

In all cases except IH1, our definitions (1.32) of N can be expressed in terms of $\langle wT \rangle$ alone. One might wonder whether N in the IH1 case would be better defined as inversely proportional to $1 - 2\langle wT \rangle$ instead of to $\overline{T}_{max}$. This would be superficially identical to the IH2 and IH3 definitions of N, provided the latter are expressed in terms of $\langle wT \rangle$. However, the IH1 experiments discussed in Sect. 3.2 confirm that $\overline{T}_{max}$ behaves very much like an inverse Nusselt number, while the quantity $1 - 2\langle wT \rangle$ does not. As described at the end of Sect. 1.6.2 above, $\langle wT \rangle$ in IH1 convection instead conveys the asymmetry between upward and downward heat fluxes.

1.6.4.2 The Nusselt Number $\tilde{N}$

Since $\delta\langle T \rangle$ is physically important in IH convection, it is natural to define a Nusselt-number-like quantity that is exactly related to $\delta\langle T \rangle$, rather than to $\langle wT \rangle$ or $\overline{T}_{max}$. It works well to simply define $\tilde{N}$ as inversely proportional to $\delta\langle T \rangle$,

$$\tilde{N} := \begin{cases} \dfrac{1}{12\delta\langle T \rangle} & \text{IH1} \\[2mm] \dfrac{1}{3\delta\langle T \rangle} & \text{IH2, IH3.} \end{cases} \tag{1.33}$$

In the IH2 and IH3 cases, this definition could be interpreted as

$$\tilde{N} = \frac{\langle zJ \rangle}{\langle zJ_{cond} \rangle} \qquad \text{IH2, IH3,} \tag{1.34}$$

which is like the expression (1.31) for N with averages weighted proportionally to height. We do not define $\tilde{N}$ for RB convection, although the mean temperature in those cases merits attention also, as discussed in Sect. 1.6.3.

Table 1.4 Proven R-dependent bounds on N and $\tilde{N}$ that hold at large R. The constants c differ between cases. These are re-expressions of the bounds on $\langle wT \rangle$ and $\delta\langle T \rangle$ shown in Table 1.3

	R-dependent bound on $\langle wT \rangle$	R-dependent bound on $\delta\langle T \rangle$
RB1	$N \leq cR^{1/2}$	None
RB2, RB3	$N \leq cR^{1/3}$	None
IH1, IH2, IH3	None	$\tilde{N} \leq cR^{1/3}$

1.6.4.3   Basic Properties

In almost all cases it has been proven that $N \geq 1$ and $\tilde{N} \geq 1$, with equality holding only in the static states. These facts follow from the uniform bounds on $\langle wT \rangle$ and $\delta\langle T \rangle$ discussed in Sect. 1.6.3.1. It remains to be proven that $N \geq 1$ in the IH1 case, which would be true if $\overline{T}_{\max}$ could not exceed its static value of 1/8. In turbulent convection, it is typically expected that $N \to \infty$ and $\tilde{N} \to \infty$ as $R \to \infty$. In the IH cases, this is tantamount to $\overline{T}_{\max} \to 0$ or $\delta\langle T \rangle \to 0$. Such limiting behavior has not been proven but is supported by the experimental results described in Chap. 3.

Table 1.4 summarizes the R-dependent bounds that are known for N and $\tilde{N}$. These are simply restatements of the bounds on $\langle wT \rangle$ and $\delta\langle T \rangle$ given above in Table 1.3. In RB convection, the upper bounds on N are equivalent to upper bounds on $\langle wT \rangle$. In IH convection, the upper bounds on $\tilde{N}$ are equivalent to lower bounds on $\delta\langle T \rangle$. Upper bounds on N have not yet been proven for IH convection. In IH2 and IH3, bounds of the form $N \leq cR^{1/3}$ would follow from the upper bounds on $\langle wT \rangle$ that we have conjectured in Sect. 1.6.3.4. A bound of the same form for IH1 would require showing that $\overline{T}_{\max}$ decays no faster than $R^{-1/3}$. The quantity $\overline{T}_{\max}$ seems harder to access mathematically than volume averages like $\delta\langle T \rangle$ and $\langle wT \rangle$, which arise naturally in integral relations.

1.6.5   Diagnostic Rayleigh Numbers

The primary purpose of defining N and $\tilde{N}$ as we have is to identify similarities between the various configurations. The bounds in Table 1.4 suggest that we have almost succeeded, but the RB1 exponent is 1/2, while the others are 1/3. To bring the various cases completely into alignment, we must speak of the dependence of N and $\tilde{N}$ on *diagnostic* Rayleigh numbers, Ra and $\widetilde{Ra}$, instead of on the control Rayleigh number, R. These diagnostic parameters can be written simply as

$$Ra := \begin{cases} R & \text{RB1} \\ R/N & \text{RB2, RB3, IH1, IH2, IH3} \end{cases} \tag{1.35}$$

$$\widetilde{Ra} := R/\tilde{N} \qquad \text{IH1, IH2, IH3.} \tag{1.36}$$

In terms of these variables, the RB bounds in Table 1.4 all take the form $N \leq cRa^{1/2}$, and the IH bounds take the form $\tilde{N} \leq c\widetilde{Ra}^{1/2}$. Moreover, the analogies brought out by considering N and Ra (or $\tilde{N}$ and $\widetilde{Ra}$) are not limited to bounds; similarities emerge also in experimental data [25, 53] and heuristic scaling arguments [19].

The different definitions of R, Ra, and $\widetilde{Ra}$ can be viewed as differences in the temperature scale used to define a Rayleigh number. The dimensional temperature scales Δ used to define R in Sect. 1.3 are characteristic of the static states, whereas Ra and $\widetilde{Ra}$ effectively replace Δ with temperature scales of the flowing fluid, Δ_{Ra} and $\Delta_{\widetilde{Ra}}$. In IH1, the temperature scale of Ra is the maximum horizontally averaged temperature in the flowing fluid. In the other five cases it is the mean temperature difference between the boundaries in the flowing fluid. That is,

$$\Delta_{Ra} = \begin{cases} \Delta & \text{RB1} \\ \delta\overline{T}\Delta & \text{RB2, RB3} \\ 8\overline{T}_{\max}\Delta & \text{IH1} \\ 2\delta\overline{T}\Delta & \text{IH2, IH3.} \end{cases} \tag{1.37}$$

Replacing Δ with Δ_{Ra} in the definition (1.10) of R leads to the above definition (1.35) of Ra. In the IH cases, the temperature scale of $\widetilde{Ra}$ is the volume-averaged temperature of the flowing fluid,

$$\Delta_{\widetilde{Ra}} = \begin{cases} 12\,\delta\langle T\rangle\Delta & \text{IH1} \\ 3\,\delta\langle T\rangle\Delta & \text{IH2, IH3.} \end{cases} \tag{1.38}$$

Replacing Δ with $\Delta_{\widetilde{Ra}}$ in the definition (1.10) of R leads to the above definition (1.36) of $\widetilde{Ra}$.

Appendix

In this appendix we prove various bounds on $\langle wT\rangle$ and $\delta\langle T\rangle$ that are uniform in the parameters, R and Pr. Most of these results are standard, but it is difficult to trace their origins, and we do not try.

Extremum Principles

In each configuration with $T = 0$ on a boundary, there holds a minimum principle giving pointwise, instantaneous lower bounds on $T(\mathbf{x},t)$. For simplicity we assume that solutions to the Boussinesq equations exist and remain smooth. If $T(\mathbf{x},t)$

ever achieves a local minimum on the interior, then at that point $\mathbf{u} \cdot \nabla T = 0$ and $\nabla^2 T \geq 0$, and so $\partial_t T \geq 0$. Thus, if the interior is initially warmer than the fixed boundary temperature of zero, it remains warmer for all time. Even if part of the interior is initially cooler than the boundary, it will be warmer than the boundary at large times. In the RB1 case, an analogous maximum principle holds relative to the warmer boundary, on which $T = 1$. For all $\mathbf{x}$ on the interior and sufficiently large t,

$$T(\mathbf{x},t) > 0 \qquad \text{RB1, RB3, IH1, IH3} \tag{1.39}$$

$$T(\mathbf{x},t) < 1 \qquad \text{RB1.} \tag{1.40}$$

In the RB2 and IH2 configurations, where fixed-flux conditions are imposed on both boundaries, neither maximum nor minimum principles hold point-wise.

Mean Convective Transport

Uniform bounds on the mean convective flux, $\langle wT \rangle$, in our RB and IH configurations are summarized in Tables 1.1 and 1.2. Many of these bounds follow from the power integrals (2.37)–(2.38), which are stated below for convenience.

$$\langle |\nabla \mathbf{u}|^2 \rangle = R \langle wT \rangle$$

$$\langle |\nabla T|^2 \rangle = \begin{cases} 1 + \langle wT \rangle & \text{RB1} \\ \delta \overline{T} = 1 - \langle wT \rangle & \text{RB2, RB3} \\ \delta \langle T \rangle & \text{IH1, IH2, IH3.} \end{cases}$$

In all six configurations, the $\mathbf{u}$ power integral implies $\langle wT \rangle \geq 0$. Since $\langle |\nabla \mathbf{u}|^2 \rangle > 0$ if convection persists, $\langle wT \rangle$ saturates its lower bound of zero if and only if the system approaches the static state as $t \to \infty$.

The uniform upper bounds on $\langle wT \rangle$ given in Tables 1.1 and 1.2 follow in most cases from lower bounds on $\delta \overline{T}$. In RB1 there is no upper bound on $\langle wT \rangle$. We get $\delta \overline{T} > 0$ from the T power integral in RB2 and RB3 and from the minimum principle in IH3. This lower bound on $\delta \overline{T}$ gives $\langle wT \rangle < 1$ in RB2 and RB3, where $\langle wT \rangle + \delta \overline{T} = 1$, and it gives $\langle wT \rangle < 1/2$ in IH1 and IH3, where $\langle wT \rangle + \delta \overline{T} = 1/2$. These upper bounds on $\langle wT \rangle$ are probably approached by certain solutions, including the turbulent attractors, as $R \to \infty$. If so, they are tight among uniform bounds. In IH1, the upper bound $\langle wT \rangle < 1/2$ follows from the minimum principle and relation (1.27).

It is likely that $\delta \overline{T} > 0$ in IH2 also, but we settle for the cruder estimate $\delta \overline{T} > -1/\sqrt{3}$, derived as follows.

$$|\delta\overline{T}| = |\langle \partial_z T \rangle|$$
$$\leq \langle |\partial_z T| \rangle$$
$$\leq \langle |\partial_z T|^2 \rangle^{1/2}$$
$$\leq \langle |\nabla T|^2 \rangle^{1/2}$$
$$\leq \delta\langle T \rangle^{1/2}$$
$$\leq \frac{1}{\sqrt{3}}.$$

The third line above follows from the Cauchy–Schwarz inequality, the fifth line follows from the T power integral, and the last line follows from the bound $\delta\langle T \rangle \leq 1/3$ that is proven below. The inequality is in fact strict since $\delta\langle T \rangle < 1/3$, except in the static state. Since $\langle wT \rangle + \delta\overline{T} = 1/2$ in IH2, the lower bound on $\delta\overline{T}$ gives $\langle wT \rangle < \frac{1}{2} + \frac{1}{\sqrt{3}}$.

In the IH1 configuration, we have also discussed $\overline{T}_{max}$, the maximum value that $\overline{T}(z)$ attains. For this quantity, the lower bound $\overline{T}_{max} > 0$ follows from the minimum principle, and the upper bound $\overline{T}_{max} < 1/\sqrt{3}$ follows from a calculation very similar to the one in the previous paragraph. This upper bound is likely not tight; it might well be that $\overline{T}_{max}$ never exceeds its static value of $1/8$.

Mean Temperature

Uniform bounds on the mean fluid temperature relative to that of the top boundary, $\delta\langle T \rangle$, are summarized in Tables 1.1 and 1.2. In the IH cases, the lower bounds $\delta\langle T \rangle > 0$ follow from the T power integral. The upper bounds are proven by integrating z^2 against the T equation (1.8), and using (1.26) in the IH1 case, to find

$$\delta\langle T \rangle = \begin{cases} \frac{1}{12} - \langle (z - \frac{1}{2})wT \rangle & \text{IH1} \\ \frac{1}{3} - \langle zwT \rangle & \text{IH2, IH3.} \end{cases} \tag{1.41}$$

Incompressibility gives $\overline{w} = 0$ and thus $\langle wT \rangle = \langle w\theta \rangle$ and $\langle zwT \rangle = \langle zw\theta \rangle$, where θ is the deviation of T from its static profile. Integrating the temperature fluctuation equation (2.3) against θ gives $\langle (z - \frac{1}{2})w\theta \rangle = \langle |\nabla\theta|^2 \rangle \geq 0$ in IH1, and likewise for $\langle zw\theta \rangle$ in the other two cases. Therefore, $\delta\langle T \rangle \leq 1/12$ in IH1, and $\delta\langle T \rangle \leq 1/3$ in IH2 and IH3.

In the RB cases, none of the uniform bounds on $\delta\langle T \rangle$ are likely to be tight. The extremum principles give $0 < \delta\langle T \rangle < 1$ in RB1 and $0 < \delta\langle T \rangle$ in RB3. The upper bound for RB3 and the upper and lower bounds for RB2 follow from the inequality $|\delta\langle T \rangle| \leq 1/\sqrt{3}$ that is derived below.

$$|\delta\langle T\rangle| = |\langle z\partial_z T\rangle|$$

$$\leq \langle |z\partial_z T|\rangle$$

$$\leq \langle z^2\rangle^{1/2}\langle \partial_z T^2\rangle^{1/2}$$

$$\leq \tfrac{1}{\sqrt{3}}\langle |\nabla T|^2\rangle^{1/2}$$

$$\leq \tfrac{1}{\sqrt{3}}(1 - \langle wT\rangle)^{1/2}$$

$$\leq \tfrac{1}{\sqrt{3}}.$$

The first line of the derivation follows from integration by parts, the third line follows from the Cauchy–Schwarz inequality, the fifth line follows from the T power integral, and the last line follows from the bound $\langle wT\rangle \geq 0$ that is proven above.

References

1. Ahlers, G., Grossmann, S., Lohse, D.: Heat transfer and large scale dynamics in turbulent Rayleigh-Bénard convection. Rev. Mod. Phys. **81**(2), 503–537 (2009)
2. Ames, K.A., Straughan, B.: Penetrative convection in fluid layers with internal heat sources. Acta Mech. **85**, 137–148 (1990)
3. Asfia, F.J., Dhir, V.K.: An experimental study of natural convection in a volumetrically heated spherical pool bounded on top with a rigid wall. Nucl. Eng. Des. **163**(3), 333–348 (1996)
4. Aurnou, J., Andreadis, S., Zhu, L., Olson, P.: Experiments on convection in Earth's core tangent cylinder. Earth Planet. Sci. Lett. **212**, 119–134 (2003)
5. Avsec, D.: Tourbillons thermoconvectifs tans l'air. Application à la météorologie. Ph.D. thesis, Université de Paris (1939)
6. Calkins, M.A., Noir, J., Eldredge, J.D., Aurnou, J.M.: The effects of boundary topography on convection in Earth's core. Geophys. J. Int. **189**, 799–814 (2012)
7. Cardin, P., Olson, P.: Chaotic thermal convection in a rapidly rotating spherical shell: consequences for flow in the outer core. Phys. Earth Planet. Inter. **82**, 235–259 (1994)
8. Chandrasekhar, S.: Hydrodynamic and Hydromagnetic Stability. Dover, New York (1981)
9. Chapman, C.J., Childress, S., Proctor, M.R.E.: Long wavelength thermal convection between non-conducting boundaries. Earth Planet. Sci. Lett. **51**, 362–369 (1980)
10. Cheung, F.B., Chawla, T.C.: Complex heat transfer processes in heat-generating horizontal fluid layers. In: Annual Re view of Numerical Fluid Mechanics and Heat Transfer, vol. 1, pp. 403–448. Hemisphere, New York (1987)
11. Chillà, F., Schumacher, J.: New perspectives in turbulent Rayleigh-Bénard convection. Eur. Phys. J. E **35**(7), 1–25 (2012)
12. Choffrut, A., Nobili, C., Otto, F.: Upper bounds on Nusselt number at finite Prandtl number. arXiv:1412.4812v1 (2014)
13. Emanuel, K.A.: Atmospheric Convection. Oxford University Press, Oxford (1994)
14. Fearn, D.R., Loper, D.E.: Compositional convection and stratification of Earth's core. Nature **289**, 393–394 (1981)
15. Featherstone, N.A., Browning, M.K., Brun, A.S., Toomre, J.: Effects of fossil magnetic fields on convective core dynamos in A-type stars. Astrophys. J. **705**, 1000–1018 (2009)
16. Galdi, G.P., Straughan, B.: Exchange of stabilities, symmetry, and nonlinear stability. Arch. Ration. Mech. Anal. **89**(3), 211–228 (1985)

17. Gastine, T., Yadav, R.K., Morin, J., Reiners, A., Wicht, J.: From solar-like to antisolar differential rotation in cool stars. Mon. Not. R. Astron. Soc. Lett. **438**, 76–80 (2014)
18. Getling, A.V.: Rayleigh-Bénard Convection: Structures and Dynamics. World Scientific Publishing Co, Singapore (1998)
19. Goluskin, D., Spiegel, E.A.: Convection driven by internal heating. Phys. Lett. A **377**(1-2), 83–92 (2012)
20. Goluskin, D., Johnston, H., Flierl, G.R., Spiegel, E.A.: Convectively driven shear and decreased heat flux. J. Fluid Mech. **759**, 360–385 (2014)
21. Grötzbach, G., Wörner, M.: Direct numerical and large eddy simulations in nuclear applications. Int. J. Heat Fluid Flow **20**(3), 222–240 (1999)
22. Heimpel, M., Aurnou, J., Wicht, J.: Simulation of equatorial and high-latitude jets on Jupiter in a deep convection model. Nature **438**, 193–6 (2005)
23. Houseman, G.: The dependence of convection planform on mode of heating. Nature **332**, 346–349 (1988)
24. Hurle, D.T.J., Jakeman, E., Pike, E.R.: On the solution of the Bénard problem with boundaries of finite conductivity. Proc. R. Soc. A **296**(1447), 469–475 (1967)
25. Johnston, H., Doering, C.R.: Comparison of turbulent thermal convection between conditions of constant temperature and constant flux. Phys. Rev. Lett. **102**(6), 064501 (2009)
26. Jones, C.A.: A dynamo model of Jupiter's magnetic field. Icarus **241**, 148–159 (2014)
27. Joseph, D.D., Shir, C.C.: Subcritical convective instability: part 1. Fluid layers. J. Fluid Mech. **26**(4), 753–768 (1966)
28. Kaspi, Y., Flierl, G.R., Showman, A.P.: The deep wind structure of the giant planets: results from an anelastic general circulation model. Icarus **202**(2), 525–542 (2009)
29. Kippenhahn, R., Weigert, A.: Stellar Structure and Evolution. Springer, New York (1994)
30. Kolmychkov, V.V., Mazhorova, O.S., Shcheritsa, O.V.: Numerical study of convection near the stability threshold in a square box with internal heat generation. Phys. Lett. A **377**, 2111–2117 (2013)
31. Kulacki, F.A., Richards, D.E.: Natural convection in plane layers and cavities with volumetric energy sources. In: Natural Convection: Fundamentals and Applications, pp. 179–254. Hemisphere, New York (1985)
32. Lohse, D., Xia, K.Q.: Small-scale properties of turbulent Rayleigh-Bénard convection. Annu. Rev. Fluid Mech. **42**(1), 335–364 (2010)
33. Marshall, J., Schott, F.: Open-ocean convection: observations, theory, and models. Rev. Geophys. **37**, 1–64 (1999)
34. Nourgaliev, R.R., Dinh, T.N., Sehgal, B.R.: Effect of fluid Prandtl number on heat transfer characteristics in internally heated liquid pools with Rayleigh numbers up to 10^{12}. Nucl. Eng. Des. **169**, 165–184 (1997)
35. Oberbeck, A.: Ueber die wärmeleitung der flüssigkeiten bei berücksichtigung der strömungen infolge von temperaturdifferenzen. Ann. Phys. **243**(6), 271–292 (1879)
36. Otero, J., Wittenberg, R.W., Worthing, R.A., Doering, C.R.: Bounds on Rayleigh-Bénard convection with an imposed heat flux. J. Fluid Mech. **473**, 191–199 (2002)
37. Rajagopal, K.R., Ruzicka, M., Srinivasa, A.R.: On the Oberbeck-Boussinesq approximation. Math. Model. Methods Appl. Sci. **6**(8), 1157–1167 (1996)
38. Rayleigh, Lord: On convection currents in a horizontal layer of fluid, when the higher temperature is on the under side. Philos. Mag. **32**(192), 529–546 (1916)
39. Schubert, G., Turcotte, D.L., Olson, P.: Mantle Convection in the Earth and Planets. Cambridge University Press, Cambridge (2001)
40. Siggia, E.D.: High Rayleigh number convection. Annu. Rev. Fluid Mech. **26**, 137–168 (1994)
41. Soderlund, K.M., Schmidt, B.E., Wicht, J., Blankenship, D.D.: Ocean-driven heating of Europa's icy shell at low latitudes. Nat. Geosci. **7**(12), 16–19 (2014)
42. Sotin, C., Labrosse, S.: Three-dimensional thermal convection in an iso-viscous, infinite Prandtl number fluid heated from within and from below: applications to the transfer of heat through planetary mantles. Phys. Earth Planet. Inter. **112**, 171–190 (1999)

43. Sparrow, E.M., Goldstein, R.J., Jonsson, V.K.: Thermal instability in a horizontal fluid layer: effect of boundary conditions and non-linear temperature profile. J. Fluid Mech. **18**(04), 513–528 (1964)
44. Spiegel, E.A.: Thermal turbulence at very small Prandtl number. J. Geophys. Res. **67**(8), 3063–3070 (1962)
45. Spiegel, E.A.: Convection in stars I. Basic Boussinesq convection. Annu. Rev. Astron. Astrophys. **9**, 323–352 (1971)
46. Stern, M.E. (ed.): Ocean Circulation Physics. Academic Press, New York (1975)
47. Storey, B.D., Zaltzman, B., Rubinstein, I.: Bulk electroconvective instability at high Péclet numbers. Phys. Rev. E **76**(4), 041501 (2007)
48. Straughan, B.: Triply resonant penetrative convection. Proc. R. Soc. A **468**, 3804–3823 (2012)
49. Thirlby, R.: Convection in an internally heated layer. J. Fluid Mech. **44**(04), 673–693 (1970)
50. Thompson, P.A.: Compressible Fluid Dynamics. McGraw-Hill Inc, New York (1972)
51. van der Poel, E.P., Ostilla-Mónico, R., Verzicco, R., Lohse, D.: Effect of velocity boundary conditions on the heat transfer and flow topology in two-dimensional Rayleigh-Bénard convection. Phys. Rev. E **90**(1), 013017 (2014)
52. Vel'tishchev, N.F.: Convection in a horizontal fluid layer with a uniform internal heat source. Fluid Dyn. **39**(2), 189–197 (2004)
53. Verzicco, R., Sreenivasan, K.R.: A comparison of turbulent thermal convection between conditions of constant temperature and constant heat flux. J. Fluid Mech. **595**, 203–219 (2008)
54. Wittenberg, R.W.: Bounds on Rayleigh-Bénard convection with imperfectly conducting plates. J. Fluid Mech. **665**, 158–198 (2010)

Chapter 2
Stabilities and Bounds

Abstract Three configurations of internally heated convection and three configurations of Rayleigh–Bénard convection are analyzed mathematically. The standard
methods of determining the linear and energy stability thresholds of the static states
are explained. These analyses yield Rayleigh numbers above which the static states
are linearly unstable and Rayleigh numbers below which they are globally stable.
The resulting thresholds are reported with high precision for various boundary
conditions on the velocity. Exact analytical values, calculated by long-wavelength
asymptotics, are given for configurations with heat fluxes fixed at both boundaries.
It is then explained how the background method is used to prove lower bounds on
the mean fluid temperature in all three internally heated configurations, including
one for which no bound has been reported previously. In every configuration, these
bounds guarantee that the mean temperature of the fluid grows with the rate of
volumetric heating, H, no slower than $H^{2/3}$. The bounds are compared with upper
bounds on the Nusselt number in RB convection.

The preceding chapter has defined six convective configurations and described, for
each case, how various integral quantities characterize bulk heat transport. The
quantities of central importance include the mean vertical transport by convection,
$\langle wT \rangle$, in RB and IH convection and the mean temperature of the fluid relative to that
of the top boundary, $\delta\langle T \rangle$, in IH convection. We would like to predict the values
that these quantities assume for various Rayleigh numbers, Prandtl numbers, and
confining geometries. This is equivalent to predicting the parameter-dependence
of the Nusselt numbers we have defined; N is determined by $\overline{T}_{max}$ in IH1 and
by $\langle wT \rangle$ in the other five cases, and $\tilde{N}$ is determined by $\delta\langle T \rangle$ in the three IH
cases. The task before us is very difficult in general and would require a greatly
improved understanding of fluid turbulence, so we are limited to partial results. The
present chapter presents facts about heat transport that can be determined purely by
mathematical analysis of the Boussinesq equations, while the next chapter addresses
simulations and laboratory experiments.

There are two main ways of studying heat transport analytically: examining
simple particular solutions that can be written down exactly or asymptotically, and
deriving bounds on integral quantities that apply to all solutions. The first method
yields much stronger results but is useful only at small Rayleigh numbers, where

© Springer International Publishing Switzerland 2016   27
D. Goluskin, *Internally Heated Convection and Rayleigh-Bénard Convection*,
SpringerBriefs in Applied Sciences and Technology,
DOI 10.1007/978-3-319-23941-5_2

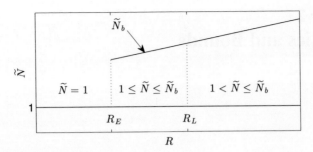

Fig. 2.1 Schematic of what this chapter's analytical results say about the dependence of $\tilde{N}$ on R in IH convection. The numerical values of R_E, R_L, and $\tilde{N}_b(R)$ vary between configurations. In RB convection, the analogous diagram for N lacks a middle region since $R_E = R_L$

the system either remains static or assumes a simple flow. Here we consider only the static states. Heat transport in static states is purely conductive and easy to understand, so the main task is to determine the parameters at which such states are stable. To this end, we can find a Rayleigh number, R_L, above which a static state is linearly unstable, and a Rayleigh number, R_E, below which we can prove that it is the unique globally stable state. These results work together with the second method of analysis—bounding N or $\tilde{N}$ above by functions $N_b(R)$ or $\tilde{N}_b(R)$—to constrain the dependence of Nusselt numbers on R. Sections 2.1–2.3 outline the calculations and values of R_L, R_E, and $\tilde{N}_b(R)$, respectively, for the various configurations.

The schematic of Fig. 2.1 shows how R_L, R_E, and $\tilde{N}_b(R)$ combine to give some knowledge of $\tilde{N}$ in IH convection. In the lowest-R regime, where $R < R_E$, we know that $\tilde{N} = 1$. This is because the system asymptotically approaches the static state, so its Nusselt number, which we have defined as an infinite-time limit, must be that of the static state. In the subcritical regime, where $R_E < R < R_L$, the static state is linearly stable, but sustained flow might also be possible, so all we can say is that $1 \leq \tilde{N} \leq \tilde{N}_b(R)$ in this regime. In the larger-R regime where $R_L < R$, the static state is linearly unstable, so any physically realizable state must have sustained flow and thus a Nusselt number strictly greater than unity. That is, $1 < \tilde{N} \leq \tilde{N}_b(R)$ for attracting states, although $\tilde{N} = 1$ remains possible if unstable states are included. A schematic like Fig. 2.1 for the other Nusselt number, N, in IH convection would lack the upper bound since R-dependent bounds on $\langle wT \rangle$ have not yet been proven (cf. Sect. 1.6.3.4).

Figure 2.1 represents a scenario where R_E is strictly smaller than R_L. In RB convection there is no subcritical regime since $R_E = R_L$. This allows for asymptotic solutions in the weakly supercritical regime, giving more precise expressions for Nusselt numbers there. In IH convection, on the other hand, subcritical convection is not ruled out because $R_E < R_L$. Stronger methods of analysis may be able to prove stability thresholds larger than R_E, but not necessarily as large as R_L. Subcritical convection is indeed possible in IH1 [42] and IH3 [3, 35, 43]. In IH2, the possibility of subcritical convection remains open.

Table 2.1 For no-slip boundary conditions: A Rayleigh number above which the static state is linearly unstable (R_L), a Rayleigh number below which the static state is globally stable (R_E), and upper bounds (N_b and $\tilde{N}_b$) on the Nusselt numbers N and $\tilde{N}$ that are valid for asymptotically large R. References for these values are given throughout the chapter. In the IH cases, R-dependent bounds on N have not been proven. In the RB cases, $\tilde{N}$ is not defined

	RB1	RB2	RB3	IH1	IH2	IH3
R_E	1707.76	720	1295.78	26,926.6	1429.86	2737.16
R_L	1707.76	720	1295.78	37,325.2	1440	2772.27
N_b	$0.027\,Ra^{1/2}$	$0.28\,Ra^{1/2}$	$0.28\,Ra^{1/2}$	None	None	None
$\tilde{N}_b$				$0.025\,\widetilde{Ra}^{1/2}$	$0.13\,\widetilde{Ra}^{1/2}$	$0.094\,\widetilde{Ra}^{1/2}$

For no-slip conditions on the velocity, Fig. 2.1 is made concrete in each configuration by Table 2.1, which gives values for R_L, R_E, $N_b(R)$, and $\tilde{N}_b(R)$. The bounds have been simplified by assuming that R is asymptotically large, and they are stated in terms of the diagnostic Rayleigh numbers, Ra and $\widetilde{Ra}$, that we have defined in Sect. 1.6.5. (Recall that Ra equals R in RB1 but equals R/N in the other five cases, and that $\widetilde{Ra} = R/\tilde{N}$ in IH convection.) The similarities between the various bounds are evident. The present chapter explains how the values in Table 2.1 are calculated and gives values for some other boundary conditions on the velocity.

The linear and nonlinear stability analyses that we apply to the static state can be applied to other particular solutions as well. Such analyses must be carried out asymptotically or numerically, however, since none of the finite-amplitude particular solutions can be expressed in closed form. The weakly nonlinear regime of IH convection has been theoretically examined in a few studies [34, 35, 40, 43]. Such analyses of particular solutions reveal much about bifurcations and pattern formation, but they do not yield robust information about heat transport. This is because the results often depend strongly on geometry, and also because each particular solution typically is stable over only a narrow range of parameters.

Sections 2.1 and 2.2 address each static state's linear and nonlinear stability, respectively. Section 2.3 outlines a proof of R-dependent lower bounds on the mean temperature for all three IH configurations. These bounds, which amount to upper bounds on $\tilde{N}$, are then compared with upper bounds on N in RB convection.

The results laid out in this chapter constitute most of what can be deduced mathematically about N and $\tilde{N}$. These results are rather meager in that they tell us neither the actual values assumed in the ranges $1 \leq N \leq N_b(R)$ and $1 \leq \tilde{N} \leq \tilde{N}_b(R)$, nor how the Prandtl number and geometry affect these values. Such questions must wait until the next chapter because substantial answers, so far, come only from simulations and laboratory experiments.

2.1   Linear Instability of Static States

In each RB and IH configuration, we can find a Rayleigh number, R_L, above which
the static state is linearly unstable. The Prandtl number of the fluid does not affect
this threshold. In most cases it has been proven that the linear instability is stationary,
meaning that the non-static states that bifurcate at the point of instability are steady,
rather than time-dependent. The method of calculating R_L is similar in every case
and is well known from the study of the canonical RB1 system. We outline this
methods here and give references for further details.

2.1.1   Linear Stability Eigenproblem

We want to study the stability of the static state, wherein $\mathbf{u} = \mathbf{0}$ and $T = T_{st}(z)$ for
the various $T_{st}(z)$ profiles given in expression (1.12). It is convenient to decompose
the temperature field into its static part and a fluctuation, θ,

$$T(\mathbf{x},t) = T_{st}(z) + \theta(\mathbf{x},t).$$

Since $\mathbf{u}$ and T evolve according to the Boussinesq equations (1.6)–(1.8), fluctuations
evolve according to

$$\nabla \cdot \mathbf{u} = 0 \tag{2.1}$$

$$\partial_t \mathbf{u} + \mathbf{u} \cdot \nabla \mathbf{u} = -\nabla p + Pr \nabla^2 \mathbf{u} + Pr R \theta \hat{\mathbf{z}} \tag{2.2}$$

$$\partial_t \theta + \mathbf{u} \cdot \nabla \theta = \nabla^2 \theta - T'_{st} w, \tag{2.3}$$

where the prime denotes $\frac{d}{dz}$. The static state enters the fluctuation equations only
through its gradient, T'_{st}, reflecting the fact that Boussinesq dynamics are affected
only by relative temperature differences, not absolute temperatures. This gradient is
constant in RB convection but varies linearly in IH convection when the heating is
uniform:

$$T'_{st}(z) = \begin{cases} -1 & \text{RB1, RB2, RB3} \\ -z + \frac{1}{2} & \text{IH1} \\ -z & \text{IH2, IH3,} \end{cases} \tag{2.4}$$

where we recall that $0 \leq z \leq 1$. The boundary conditions on θ are the homogenous
analogs of the conditions on T:

$$\text{RB1, IH1:} \qquad \theta|_{z=0}, \quad \theta|_{z=1} = 0 \tag{2.5}$$

$$\text{RB2, IH2:} \quad \partial_z\theta|_{z=0}, \ \partial_z\theta|_{z=1} = 0 \tag{2.6}$$

$$\text{RB3, IH3:} \quad \partial_z\theta|_{z=0}, \quad \theta|_{z=1} = 0. \tag{2.7}$$

The fluctuation dynamics of RB1 and IH1 are distinguished only by differing $\overline{T}'_{st}(z)$. The same is true of RB2 and IH2, and of RB3 and IH3.

We study the stability of the zero solution of the fluctuation equations (2.1)–(2.3), which is equivalent to the stability of the static state. The nonlinear terms in the fluctuation equations can be neglected when finding linear stability thresholds. As is standard [4], we find a closed pair of equations governing the linear evolution of w and θ by taking $\hat{\mathbf{z}} \cdot \nabla \times \nabla \times (2.2)$. Omitting time derivatives gives equations for the marginally stable states that are stationary, meaning they do not vary in time:

$$\nabla^4 w = -R\nabla_H^2\theta \tag{2.8}$$

$$\nabla^2\theta = T'_{st}w, \tag{2.9}$$

where $\nabla_H^2 := \partial_x^2 + \partial_y^2$ is the horizontal Laplacian operator. The validity of considering only stationary instabilities is discussed at the end of this subsection.

The Rayleigh number, R_L, at which the static state becomes linearly unstable is the smallest R for which there is a marginally stable state—that is, the smallest R for which Eqs. (2.8)–(2.9) have a nonzero solution. This is a (generalized) eigenproblem whose spectrum of eigenvalues is continuous and bounded below. Assuming there are no horizontal boundaries, we can Fourier transform the eigenproblem in x and y, decomposing it into an independent eigenproblem for each horizontal wavevector (k_x, k_y), where k_x and k_y are real. If the horizontal periods of a mode are L_x and L_y, then $k_x = 2\pi/L_x$ and $k_y = 2\pi/L_y$. The resulting decomposed eigenproblems take the form [4, 33]

$$\hat{w}^{(4)} - 2k^2\hat{w}'' + k^4\hat{w} = Rk^2\hat{\theta} \tag{2.10}$$

$$\hat{\theta}'' - k^2\hat{\theta} = T'_{st}\hat{w}, \tag{2.11}$$

where $\hat{w}(z)$ and $\hat{\theta}(z)$ are complex in general, and $k^2 := k_x^2 + k_y^2$. We call k the horizontal wavenumber.

The sixth-order linear system (2.10)–(2.11) requires six boundary conditions. The conditions (2.5)–(2.7) on θ apply also to $\hat{\theta}$. The first two $\hat{w}$ conditions are that $\hat{w} = 0$ at both boundaries, and the other two depend on whether each boundary is no-slip or free-slip:

$$\text{no-slip:} \quad \hat{w}'(0), \ \hat{w}'(1) = 0 \tag{2.12}$$

$$\text{free-slip top:} \quad \hat{w}'(0), \ \hat{w}''(1) = 0 \tag{2.13}$$

$$\text{free-slip bottom:} \quad \hat{w}''(0), \ \hat{w}'(1) = 0 \tag{2.14}$$

$$\text{free-slip:} \quad \hat{w}''(0), \ \hat{w}''(1) = 0. \tag{2.15}$$

Throughout this work, we give results for all four pairs of velocity conditions when possible. In some cases, analytical bounds and experimental results are available only for no-slip boundaries, which are the most natural in the laboratory. Condition (2.13) is experimentally realizable in a container with an open top. Condition (2.14) might appear unrealizable since it describes a container with an open bottom, but it also describes dynamically equivalent systems with an open top. For instance, IH convection with an open bottom, when viewed upside down, has the same dynamics as internally *cooled* convection with an open top.

For each k^2, Eqs. (2.10)–(2.11) and their boundary conditions form a linear eigenproblem in R with a *discrete* spectrum that is easier to compute than the continuous spectrum of (2.8)–(2.9). The R_L at which the static state loses stability is the smallest eigenvalue of (2.10)–(2.11), minimized over all admissible k^2. If all horizontal wavenumbers are possible,

$$R_L = \inf_{k^2 > 0} R^{(0)}(k), \qquad (2.16)$$

where

$$R^{(0)}(k) := \min \left\{ R \mid (2.10)\text{–}(2.11) \text{ has a nonzero solution} \right\}. \qquad (2.17)$$

The definition of R_L requires an infimum rather than a minimum because the infimum sometime occurs in the limit $k^2 \to 0$, in which case no minimum is achieved. Perturbations with $k = 0$ are not admissible since a horizontally uniform $\hat{w}$ would violate incompressibility. The value (or limit) of k at which R_L occurs is called the critical wavenumber of linear instability, k_L.

Because the eigenproblem (2.8)–(2.9) is derived assuming a stationary instability, the resulting R_L is the value at which a steady state bifurcates from the static one. In cases where it is proven that all marginally stable states are indeed stationary, $R > R_L$ is not only sufficient but necessary for instability of the static state. Stationarity has been proven for all RB cases [31]. In the IH3 case it follows for free-slip boundaries from an argument of Spiegel (see footnote 4 of [44]) and for no-slip boundaries from a theorem of Herron [16]. The latter method of proof may suffice to show stationarity in the remaining IH configurations. Until that is done, we can say in those cases only that $R > R_L$ is sufficient for instability.

2.1.2 Solutions of the Linear Stability Eigenproblem

In RB convection, where $T'_{st} = -1$, the eigenfunctions solving (2.10)–(2.11) are combinations of trigonometric and hyperbolic functions. The minimum eigenvalue at a given wavenumber, $R^{(0)}(k)$, must satisfy an expression involving the eigenfunctions. This expression can be solved for $R^{(0)}(k)$, analytically in a few cases and numerically in the others [4, 33]. In IH convection, where T'_{st} varies linearly

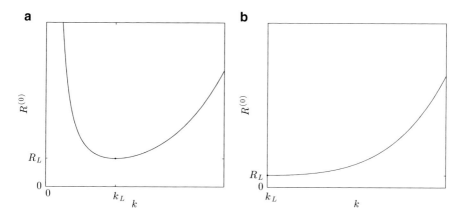

Fig. 2.2 Schematic diagrams of how the first marginally stable eigenvalue, $R^{(0)}$, depends on the horizontal wavenumber, k, in (**a**) the RB1, RB3, IH1, and IH3 cases, and in (**b**) the RB2 and IH2 cases, where heat fluxes are fixed at both boundaries

with z, the analogous approach would involve hypergeometric functions, so it is simpler to solve the eigenproblem (2.10)–(2.11) numerically. We have done this for all six configurations by the general method described in [41]: discretizing the differential operators using a spectral collocation method and computing the spectra of the resulting matrices. Our computed values of R_L agree with or add precision to the values in the literature. As explained shortly, the exact values of R_L in the RB2 and IH2 cases can be calculated also by asymptotic expansion.

The $R^{(0)}(k)$ curve can assume one of two qualitative shapes in our models, depending on the thermal boundary conditions. Figure 2.2a shows what the $R^{(0)}(k)$ curves look like in the four cases where the temperature is fixed at one or both boundaries: $R^{(0)}$ is minimized at a finite k and grows unboundedly both as $k \to 0$ and as $k \to \infty$. Figure 2.2b shows what the $R^{(0)}(k)$ curves look like when the temperature flux is fixed at both boundaries: $R^{(0)}$ approaches its infimum as $k \to 0$ and grows unboundedly as $k \to \infty$.

For all of our RB and IH configurations and all four pairs of velocity conditions (2.12)–(2.15), Table 2.2 gives the smallest Rayleigh number, R_L, at which the static state undergoes a stationary, horizontally periodic instability, along with the instability's horizontal wavenumber, k_L. Most of these values have been known for a long time. The RB1 case was analyzed first, with free-slip boundaries in Rayleigh's seminal analysis of 1916 [33] and then with other velocity conditions [20, 26, 31]. Later, values of R_L for some of our velocity conditions were reported for RB2 and RB3 [37], IH1 [11, 24, 37, 44, 45], IH3 [11, 24, 28, 34], and then IH2 [15, 19].

In most configurations, R_L is smallest when both boundaries are free-slip, largest when both boundaries are no-slip, and somewhere in between when one boundary is free-slip and the other is no-slip. The IH1 configuration provides a surprising exception: R_L is smallest when only the top is free-slip and largest when only the bottom is free-slip.

Table 2.2 Rayleigh number, R_L, above which each configuration's static state is linearly unstable, and the horizontal wavenumber, k_L, of the linear perturbation that is marginally stable at R_L. The RB2 and IH2 values are exact (cf. Sect. 2.1.3), while the other values are numerical approximations that are accurate to the precision shown

	R_L	k_L
RB1		
No-slip	1707.76	3.1163
Free-slip top	1100.65	2.6823
Free-slip bottom	1100.65	2.6823
Free-slip	657.511	2.2214
RB2		
No-slip	720	0
Free-slip top	320	0
Free-slip bottom	320	0
Free-slip	120	0
RB3		
No-slip	1295.78	2.5519
Free-slip top	816.744	2.2147
Free-slip bottom	668.998	2.0856
Free-slip	384.693	1.7576
IH1		
No-slip	37,325.2	3.9989
Free-slip top	16,669.8	3.0131
Free-slip bottom	37,949.4	4.0867
Free-slip	16,992.2	3.0277
IH2		
No-slip	1440	0
Free-slip top	576	0
Free-slip bottom	720	0
Free-slip	240	0
IH3		
No-slip	2772.27	2.6293
Free-slip top	1612.62	2.2611
Free-slip bottom	1650.55	2.1429
Free-slip	867.766	1.7897

In principle, knowing R_L is relevant to heat transport because $R > R_L$ suggests that sustained convection will occur. In confined geometries, however, not all k are admitted, and the flow can be only approximately periodic in the horizontal directions. These effects raise R_L by an amount particular to the confining geometry, and only when R exceeds this larger value is convection guaranteed.

The RB2 and IH2 configurations are special in that explicit expressions for R_L can be found analytically for any boundary conditions on the velocity. This is because the critical wavenumber is zero (cf. Table 2.2), so R_L can be calculated by long-wavelength asymptotics.

2.1.3   Long-Wavelength Asymptotics for RB2 and IH2

In the RB2 and IH2 cases, where heat fluxes are fixed at both boundaries, the infimum in the definition (2.16) of R_L occurs when $k \to 0$, so

$$R_L = \lim_{k \to 0} R^{(0)}(k). \tag{2.18}$$

It has apparently not been proven analytically that fixed-flux boundary conditions imply $k_L = 0$, so the finding must be verified on a case-by-case basis by numerically computing $R^{(0)}(k)$. This has been done for RB2 and IH2, so R_L can be found exactly by expanding the linear stability eigenproblem (2.10)–(2.11) in small k^2. (We cannot simply set $k = 0$ because the limit is singular.) The eigenmode with eigenvalue $R^{(0)}(k)$ has scaling $O(\hat\theta) = k^2 O(\hat w)$ when $k \ll 1$ [6]. We thus let $\hat w = k^2 \hat W$ and seek solutions where $\hat W$ and $\hat\theta$ are both $O(1)$. The rescaled eigenproblem is

$$\hat W^{(4)} = R\hat\theta + 2k^2 \hat W'' - k^4 \hat W \tag{2.19}$$

$$\hat\theta'' = k^2 \left(\hat\theta + T'_{\text{st}} \hat W \right). \tag{2.20}$$

The no-flux conditions on $\hat\theta$ require the vertical integral of $\hat\theta''$ to vanish, and this furnishes a consistency condition on the right-hand side of (2.20):

$$\int_0^1 \left(\hat\theta + T'_{\text{st}} \hat W \right) dz = 0. \tag{2.21}$$

Equations (2.19)–(2.21) suffice to determine R_L, but it is possible to also retain the nonlinear terms in a long-wavelength expansion of the fluctuation equations. This has been carried out for RB2 [6], IH2 [19], and some other convective models with fixed boundary fluxes [5, 8]. The simpler linear calculation we describe here is contained in these nonlinear analyses.

Equations (2.19)–(2.20) are solved asymptotically by expanding in k^2:

$$\hat W(z) = W_0(z) + k^2 W_2(z) + \cdots \tag{2.22}$$

$$\hat\theta(z) = \theta_0(z) + k^2 \theta_2(z) + \cdots \tag{2.23}$$

$$R = R_L + k^2 R_2 + \cdots . \tag{2.24}$$

The R expansion anticipates that $R_0 = R_L$—in other words, that $k_L = 0$. We and others have confirmed this for RB2 and IH2 by computing the $R^{(0)}(k)$ curves numerically.

The procedure for asymptotically solving Eqs. (2.19)–(2.20) to any order k^{2n} is entirely systematic. Assuming all lower-order terms are known, the polynomial $\theta_{2n}(z)$ is found by integrating equation (2.20) at $O(k^{2n})$, then the polynomial $W_{2n}(z)$

is found by integrating equation (2.19) at $O(k^{2n})$, and finally R_{2n} is found from the consistency condition (2.21) at $O(k^{2n})$. Since $R_L = R_0$ here, we only need to carry out these three steps at leading order to find R_L. This has been done for RB2 in [6] and for IH2 in [15, 19]. The results of the three steps are that

$$R_L = \frac{-1}{\int_0^1 T'_{\text{st}} P(z) dz} = \begin{cases} \dfrac{1}{\int_0^1 P(z) dz} & \text{RB2} \\[4mm] \dfrac{1}{\int_0^1 z P(z) dz} & \text{IH2}, \end{cases} \qquad (2.25)$$

where $P(z)$ is the unique fourth-order polynomial that has a leading coefficient of $1/24$ and satisfies the $\hat{w}$ boundary conditions. For our domain of $0 \le z \le 1$, these $P(z)$ are given in [15], for instance. The values of R_L for various velocity conditions appear in Table 2.2 above.

2.2 Energy Stability of Static States

Knowing the Rayleigh number, R_L, above which a static state is linearly unstable would be well complemented by knowing the critical Rayleigh number, R_c, below which it is the globally attracting state of the fully nonlinear dynamics. This is difficult in general, so we settle for finding a so-called *energy* Rayleigh number, R_E, that is a lower bound on R_c. Since linear instability implies nonlinear instability, we can anticipate that $R_E \le R_c \le R_L$. RB convection is special in that $R_E = R_c = R_L$ [21]. IH convection is more complicated in that $R_E < R_L$ for the largest known R_E, as depicted in Fig. 2.1. The values of R_E and R_L serve as upper and lower bounds on R_c, respectively, that hold uniformly for all Pr. The exact values of R_c can depend on Pr and are not yet known. The upper bounds on R_c can be tightened by finding particular subcritical solutions, as has been done for IH1 [42] and IH3 [35, 43], since any R at which subcritical convection persists must be larger than R_c. Proving a lower bound tighter than R_E is a more daunting challenge.

2.2.1 Lyapunov Stability and the Energy Method

The global stability of the static state is equivalent to the global stability of the zero solution to the fluctuation equations (2.1)–(2.3). The nonlinear terms in those equations that could be ignored in the linear stability analysis must now be included. The typical method of proving global stability, due to Lyapunov, requires finding a functional of the state variables that is nonnegative and whose evolution is nonpositive. That is, we must find a functional $\mathscr{L}[\mathbf{u}, \theta]$ such that

$$\mathscr{L}[\mathbf{u}, \theta] \geq 0 \tag{2.26}$$

$$\tfrac{d}{dt}\mathscr{L}[\mathbf{u}, \theta] \leq 0. \tag{2.27}$$

To show also that the static state attracts all initial conditions, it suffices for the above inequalities to be strict whenever $\mathbf{u}$ or θ is nonzero. In the convective systems we are studying, the best we can hope for is to find an $\mathscr{L}$ where (2.26) and (2.27) hold for R below some finite value, $R_{\mathscr{L}}$. That is,

$$R_{\mathscr{L}} := \sup\{R \mid \mathscr{L} \text{ satisfies } (2.26)\text{–}(2.27)\}. \tag{2.28}$$

The critical Rayleigh number R_c, is the largest R at which *any* Lyapunov functional exists,

$$R_c := \sup_{\mathscr{L}} R_{\mathscr{L}}. \tag{2.29}$$

There is no universally successful method for constructing Lyapunov functionals, let alone the optimal $\mathscr{L}$ that is valid for R up to R_c. It is even difficult to confirm that an optimal $\mathscr{L}$ is indeed optimal, except when $R_c = R_L$, as in RB convection. All we can do in general is make educated guesses for $\mathscr{L}$, determine the corresponding values of $R_{\mathscr{L}}$, and declare the largest $R_{\mathscr{L}}$ we can find to be a lower bound on R_c. In fluid dynamical systems like ours, even this guess-and-check procedure cannot be carried out for general $\mathscr{L}$ because it is too difficult to determine whether the second Lyapunov condition (2.27) holds. In most nonlinear analyses of fluid stability, this trouble is avoided by considering only a particular subset of possible Lyapunov functionals for which it is tractable to check the second Lyapunov condition. This approach is called the *energy method*.

The energy method in fluid mechanics [22, 36, 39] is a special case of Lyapunov's method. In one definition of the energy method that is neither the narrowest nor the broadest definition possible, the Lyapunov functional, which is called the energy, has two special features:

1. The energy is quadratic in the state variables.
2. The energy is conserved by the nonlinear terms of the fluctuation equations (2.2)–(2.3), meaning that these terms do not contribute to the expression for the time-evolution of the energy.

The energy method is so named because quadratic quantities are often proportional to physical energies. Here we follow Joseph [21] in considering energies of the form

$$E_\gamma[\mathbf{u}, \theta](t) := \tfrac{1}{2} \int \left(\tfrac{1}{PrR} |\mathbf{u}|^2 + \gamma\theta^2 \right) d\mathbf{x}, \tag{2.30}$$

where $\int$ denotes an instantaneous volume average. The constant $\gamma > 0$ is called a *coupling parameter*, and each positive value defines an energy that is a valid Lyapunov functional for R up to some R_{E_γ}. This value of R_{E_γ} is maximized by some optimal choice of γ, where it achieves the critical Rayleigh number of energy stability, R_E:

$$R_E := \max_{\gamma>0} \sup \left\{ R \mid E_\gamma \text{ satisfies } (2.26)\text{–}(2.27) \right\}. \tag{2.31}$$

The value of R_E is the best lower bound on R_c that we find by the energy method, though it is likely still smaller than R_c. Deriving a better lower bound on R_c would require going beyond the energy method to search over a larger class of Lyapunov functionals, and this presents technical challenges. Progress beyond the energy method has been made for a few shear flow models [7, 23] but not yet for a convective system.

2.2.2   Energy Stability Eigenproblem

The functional E_γ suffices to show that the static state is globally stable whenever it satisfies conditions (2.26)–(2.27). The first condition holds whenever all the parameters are positive, so it remains only to determine the parameters for which $\frac{d}{dt}E_\gamma \le 0$. Adding the volume averages of $\frac{1}{PrR}\mathbf{u} \cdot$ (2.2) and $\gamma\theta \times$ (2.3) and then integrating by parts gives

$$\tfrac{d}{dt}E_\gamma = -\int \left[\tfrac{1}{R}|\nabla\mathbf{u}|^2 + \gamma|\nabla\theta|^2 - \left(1 - \gamma T_{\mathrm{st}}'\right) w\theta \right] d\mathbf{x}. \tag{2.32}$$

The static state is globally attracting if the right-hand side of (2.32) is negative definite. Like the linear stability threshold, the satisfaction of this condition depends on R but not on Pr. Only static states have this feature; other solutions and their stabilities depend also on Pr.

The calculus of variations yields a necessary and sufficient condition for the righthand side of (2.32) to be negative definite. In particular, E_γ is a Lyapunov functional when R is smaller than all eigenvalues, R, of the (generalized) eigenproblem [1, 38, 39]

$$\hat{w}^{(4)} - 2k^2\hat{w}'' + k^4\hat{w} = \tfrac{1}{2}Rk^2 \left(1 - \gamma T_{\mathrm{st}}'\right) \hat{\theta} \tag{2.33}$$

$$\gamma\left(\hat{\theta}'' - k^2\hat{\theta}\right) = -\tfrac{1}{2}\left(1 - \gamma T_{\mathrm{st}}'\right) \hat{w}. \tag{2.34}$$

The boundary conditions are the same as in the linear stability eigenproblem of Sect. 2.1.2, and again $\hat{w}(z)$ and $\hat{\theta}(z)$ can be complex, and k is the horizontal wavenumber. Expression (2.31) for R_E can thus be restated as

$$R_E = \max_{\gamma>0} \inf_{k^2>0} \min \left\{ R \mid (2.33)\text{–}(2.34) \text{ has a nonzero solution} \right\}. \tag{2.35}$$

It is a special feature of the energy method, and not of Lyapunov's method in general, that the nonlinear stability analysis can be reduced to the solution of a linear eigenproblem, much like the linear stability analysis.

2.2.3   Solutions of the Energy Stability Eigenproblem

In RB convection, where $T'_{st} = -1$, there is no need to solve the energy stability eigenproblem (2.33)–(2.34) because it is identical to the linear stability eigenproblem (2.10)–(2.11), so long as the energy is defined with $\gamma = 1$. This energy is thus a valid Lyapunov functional for all R up to R_L. (The agreement of the two eigenproblems reflects the self-adjointness of the linear stability operator; see [14, 39].) We expect $\gamma = 1$ to be the optimal coupling parameter since R_E should not exceed R_L, and indeed this can be shown directly [21]. These observations justify our earlier assertion that $R_E = R_c = R_L$ in RB convection, making subcritical instability impossible.

In IH convection, R_E must be calculated by performing the double optimization of expression (2.35), which requires solving the eigenproblem (2.33)–(2.34). In all IH cases the strict inequality $R_E < R_L$ holds. Table 2.3 gives values of R_E for the various IH configurations, along with the percent differences between R_L and R_E, and the arguments, γ^* and k_E, that yield the maxima and infima in expression (2.35).

The relative magnitudes of the gaps between R_E and R_L are on the order of 1% in IH2 and IH3, where the bottom is insulating, but are much larger in IH1, where heat

Table 2.3 Rayleigh number (R_E) below which the energy method proves that each IH configuration's static state is globally attracting, the percentage of R_L by which R_E falls short of R_L, the optimal coupling parameter (γ^*) used to define the energy that is a valid Lyapunov functional for all $R < R_E$, and the horizontal wavenumber (k_E) at which the infimum in (2.35) occurs for the optimal energy. The IH2 values are numerical approximations to the exact analytical expressions (2.36). The IH1 and IH3 values are computed numerically and are accurate to the precision shown

	R_E	% below R_L	γ^*	k_E
IH1				
No-slip	26 926.6	27.9	8.8831	3.6174
Free-slip top	12 620.2	24.3	7.9626	2.9014
Free-slip bottom	24 722.8	34.9	9.1975	3.3664
Free-slip	10 618.1	37.5	8.8516	2.5498
IH2				
No-slip	1429.86	0.704	1.9720	0
Free-slip top	573.391	0.453	1.7838	0
Free-slip bottom	714.929	0.704	2.2185	0
Free-slip	239.055	0.394	1.9843	0
IH3				
No-slip	2737.16	1.27	2.0678	2.6355
Free-slip top	1594.42	1.13	1.9185	2.2661
Free-slip bottom	1624.26	1.59	2.3702	2.1512
Free-slip	855.674	1.39	2.1821	1.7958

escapes across both boundaries. We cannot say whether the larger gaps in IH1 are necessitated by subcritical solutions or are only mathematical artifacts of the optimal energies being poor approximations of the truly optimal Lyapunov functionals. The most energy-unstable wavenumber, k_E, is fairly close to k_L in IH1 and IH3, and $k_E = k_L = 0$ in IH2. The optimal coupling parameters, γ^*, are all significantly larger than unity, which is their optimal value in the RB cases.

For the IH1 and IH3 cases, we numerically computed the values of R_E, γ^*, and k_E that appear in Table 2.3 using the same spectral collocation method that we used to compute R_L. A similar energy stability analysis was carried out by Kulacki and Goldstein [25]. The values of R_E that they reported are smaller than our own, perhaps because their coupling parameters were not quite optimal.[1] An energy stability analysis was carried out more recently for the IH3 configuration with no-slip boundaries [38], and those findings agree exactly with our own.

In the IH2 case, the energy stability eigenproblem can be solved exactly using long-wavelength asymptotics, much like the linear stability eigenproblem (cf. Sect. 2.1.3). This is possible because the infimum of expression (2.35) is reached as $k^2 \to 0$, an observation that has not been proven but has been confirmed numerically [15]. The asymptotic calculations, which are detailed in [15], give the exact expressions

$$R_E = \begin{cases} 2880\left(6\sqrt{35} - 35\right) & \text{no-slip} \\ 360\left(9\sqrt{385} - 175\right) & \text{free-slip top} \\ 1440\left(6\sqrt{35} - 35\right) & \text{free-slip bottom} \\ 1440\left(8\sqrt{7} - 21\right) & \text{free-slip} \end{cases} \tag{2.36}$$

Numerical approximations of the above values appear in Table 2.3 above.

From the standpoint of scientific and engineering applications, the value of knowing R_E in IH convection is that we know convection cannot be sustained when $R < R_E$. When R lies between R_E and R_L, little is known about when convection can occur, apart from some instances of subcritical convection that have been computed in IH1 [42] and IH3 [35, 43]. This ambiguous regime between R_E and R_L is small in IH2 and IH3, and thus of not much practical importance, but it is much larger in IH1. In any event, we cannot claim to fully understand the static states until we know when subcritical convection is possible—that is, until we know the true value of R_c for every Pr. Lower bounds on R_c could be improved by looking beyond the energy method to find better Lyapunov functionals, and upper bounds could be improved by numerically computing steady states that exist in the subcritical regimes.

[1] What we call the IH1 case is designated in [25] by the parameter $Bi_0 = \infty$, and what we call the IH3 case is designated by the parameters $Bi_0 = 0$ and $Bi_1 = \infty$. Their Rayleigh numbers are converted to our scaling upon multiplication by 64.

2.3 Bounds Depending on the Rayleigh Number

Our main goal is to predict the parameter-dependence of integral quantities like $\langle wT \rangle$, $\delta\langle T \rangle$, and $\overline{T}_{\max}$. Much of this effort is equivalent to seeking the functions $N(R,Pr)$ and $\tilde{N}(R,Pr)$, where these Nusselt numbers are defined as in Tables 1.1 and 1.2. (Such functions are multivalued in general since multiple locally attracting solutions can coexist at a given set of parameters.) The stability analyses of Sects. 2.1 and 2.2 are useful because they give necessary and sufficient conditions for N and $\tilde{N}$ to equal unity. In particular, $R < R_E$ guarantees that both quantities equal unity, and $R > R_L$ guarantees that both are greater than unity. At large R, where convection is strong and complicated, exact expressions for $N(R,Pr)$ and $\tilde{N}(R,Pr)$ are not available. Instead, we seek to bound these quantities analytically.

The only parameter-dependent bounds that have been proven for RB or IH configurations can be stated as upper bounds on how quickly N or $\tilde{N}$, respectively, can grow as R is raised. Upper bounds on N have not been proven in IH convection but seem likely to hold (cf. Sect. 1.6.4). We cannot improve the lower bounds of unity since known techniques cannot distinguish realizable solutions from the unstable static states.

In this section we outline a proof of lower bounds on the mean temperature, $\delta\langle T \rangle := \langle T - \overline{T}_T \rangle$, in IH convection. (Recall that $\delta\langle T \rangle \equiv \langle T \rangle$ in IH1 and IH3 but not in IH2, and that lower bounds on $\delta\langle T \rangle$ are equivalent to upper bounds on $\tilde{N}$.) Our exposition combines existing results for IH1 [27] and IH2 [15] and a new result for IH3. The proof employs the background method [10, 12], which requires no assumptions beyond the governing equations. Like similar variational methods [2, 17, 18], the background method makes progress by relaxing the constraints on $\mathbf{u}$ and T. Instead of enforcing the full Boussinesq equations, we enforce only incompressibility, the boundary conditions, and a few integral relations that follow from the governing equations. This yields bounds that hold for an enlarged class of $\mathbf{u}$ and T that includes solutions of the Boussinesq equations.

Two main integral relations are typically enforced when the background method is applied to convection. They are called the power integrals and are derived by taking $\langle \mathbf{u} \cdot (1.7) \rangle$ and $\langle T \times (1.8) \rangle$ and integrating by parts to find

$$\langle |\nabla \mathbf{u}|^2 \rangle = R \langle wT \rangle \tag{2.37}$$

$$\langle |\nabla T|^2 \rangle = \begin{cases} 1 + \langle wT \rangle & \text{RB1} \\ \delta\overline{T} = 1 - \langle wT \rangle & \text{RB2, RB3} \\ \delta\langle T \rangle & \text{IH1, IH2, IH3.} \end{cases} \tag{2.38}$$

Time derivatives have vanished from the above relations in the infinite-time limit since the volume integrals of $|\mathbf{u}|$ and $|T|$ are bounded uniformly in time. This boundedness is proven as a by-product of the background-method analysis itself [10, 13]. The absence of Pr from the relaxed constraints on $\mathbf{u}$ and T precludes our analysis from producing bounds that depend on Pr.

In all six RB and IH configurations, the bounds that have been proven by the background method amount to bounds on the thermal dissipation, $\langle |\nabla T|^2 \rangle$, though they are often stated in terms of quantities like $\langle wT \rangle$ or $\delta \langle T \rangle$ that are related to $\langle |\nabla T|^2 \rangle$ by (2.38). The thermal dissipation is bounded above in RB1 and below in the other five cases. These results constitute upper bounds on N in RB convection and upper bounds on $\tilde{N}$ in IH convection.

2.3.1 Proof by the Background Method

We now prove for all three IH configurations that the dimensionless mean temperature, $\delta \langle T \rangle$, decays no faster than $R^{-1/3}$. This is equivalent to the *dimensional* mean temperature, $\delta \langle T \rangle \Delta$, growing with the rate of volumetric heating, H, no slower than $H^{2/3}$. We assume a no-slip top in the IH2 case but need not do so in the IH1 or IH3 cases.

To apply the background method, we decompose the temperature into a so-called background profile, $\tau(z)$, and the remaining part, $\Theta(\mathbf{x}, t)$:

$$T(\mathbf{x}, t) = \tau(z) + \Theta(\mathbf{x}, t). \tag{2.39}$$

The bound we obtain depends on the $\tau(z)$ we choose. The background profile does not generally solve the governing equations, in which case Θ does not evolve according to the fluctuation equations (2.1)–(2.3).

The $\tau(z)$ we choose must satisfy several conditions. First, it must be continuous. Second, it must satisfy the same boundary conditions as T since this lets Θ satisfy the corresponding homogenous conditions. In practice, however, only the fixed-temperature conditions on $\tau(z)$ need to be enforced. This is because fixed-flux conditions on $\tau(z)$ can be met by boundary layers whose influence vanishes as we send their thicknesses to zero [15]. These limiting bounds are the same as those reached by simply ignoring the fixed-flux conditions on $\tau(z)$, so we do the latter in our calculations. Finally, $\tau(z)$ must be chosen to make a particular quantity nonnegative, as explained below. We will see that for all admissible $\tau(z)$

$$\delta \langle T \rangle \geq 2 \langle \tau - \tau_T \rangle - \langle \tau'^2 \rangle. \tag{2.40}$$

We choose simple $\tau(z)$ that make our calculations analytically tractable, thereby yielding analytical bounds that are valid at all R. Optimizing $\tau(z)$ numerically at a given R would give a tighter bound (as in [32]), but the bound would apply only at that value of R.

To see where the inequality (2.40) comes from, and when it holds, we expand the power integral (2.38) for the IH cases using the decomposition (2.39) to find

$$\delta \langle T \rangle = \langle |\nabla T|^2 \rangle = \langle \tau'^2 \rangle + 2 \langle \tau' \Theta' \rangle + \langle |\nabla \Theta|^2 \rangle, \tag{2.41}$$

where primes denote z-derivatives. Our goal is to bound $\delta\langle T\rangle$ below. (We could equally well speak of bounding $\langle|\nabla T|^2\rangle$ below or bounding $\tilde{N}$ above.) The $\langle\tau'\Theta'\rangle$ term in the above expression is difficult to bound, so we eliminate it using a third and final integral relation. Integrating $\tau(z)$ against the temperature equation (1.8) gives the needed relation [27],

$$\langle\tau'\Theta'\rangle = \langle\tau - \tau_T\rangle - \langle\tau'^2\rangle + \langle\tau'w\Theta\rangle, \qquad (2.42)$$

where the top temperature τ_T may be nonzero only in the IH2 case. Eliminating $\langle\tau'\Theta'\rangle$ from expression (2.41) gives

$$\delta\langle T\rangle = 2\langle\tau - \tau_T\rangle - \langle\tau'^2\rangle + \langle|\nabla\Theta|^2\rangle + 2\langle\tau'w\Theta\rangle. \qquad (2.43)$$

From the above equality it follows that the lower bound (2.40) would hold if we could show $\langle|\nabla\Theta|^2\rangle + 2\langle\tau'w\Theta\rangle \geq 0$. This is an impossible task for arbitrary w, however, since the velocity enters only in the sign-indefinite term. Apparently, the temperature power integral (2.38) alone is not sufficiently constraining. We need the additional constraint of the velocity power integral (2.37), which tells us that $a\left(\frac{1}{R}\langle|\nabla\mathbf{u}|^2\rangle - \langle wT\rangle\right) = 0$ for any a. Adding this relation to (2.43) shows that the lower bound (2.40) on $\delta\langle T\rangle$ would follow from the nonnegativity of the quadratic functional

$$\mathscr{Q}[\mathbf{u},\Theta;\tau(z,R)] := \frac{a}{R}\langle|\nabla\mathbf{u}|^2\rangle + \langle|\nabla\Theta|^2\rangle + \langle(2\tau' - a)w\Theta\rangle. \qquad (2.44)$$

We must choose a $\tau(z)$ for which we can verify that $\mathscr{Q} \geq 0$ for all admissible $\mathbf{u}$ and Θ.

More generally, the background method is carried out by finding an expression, equal to the quantity to be bounded, that takes the form $\mathscr{B} + \mathscr{Q}$, where $\mathscr{B}$ is a functional of the background field alone, while $\mathscr{Q}$ depends also on the other fields. The key idea is that $\mathscr{B}$ will be a lower bound when we can show that $\mathscr{Q}$ is nonnegative (or an upper bound when we can show that $\mathscr{Q}$ is nonpositive). In the present analysis, $\mathscr{B} := 2\langle\tau - \tau_T\rangle - \langle\tau'^2\rangle$, and $\mathscr{Q}$ is as defined in (2.44).

Two objectives compete in the choice of $\tau(z)$: making the lower bound (2.40) as large as possible, and maintaining the nonnegativity of $\mathscr{Q}$ that is needed for that bound to be valid. Here, we optimize $\tau(z)$ only among profiles consisting of two linear pieces. Such profiles can all be written in the ansatz

$$\tau(z) = \begin{cases} \left[\frac{b}{\delta} + \frac{a}{2}\left(\frac{1}{\delta} - 1\right)\right](1-z) & 1-\delta \leq z \leq 1 \\ b + \frac{a}{2}z & 0 \leq z \leq 1-\delta, \end{cases} \qquad (2.45)$$

where the geometric meanings of parameters a, b, and δ are shown in Fig. 2.3. We will see that the top piece of $\tau(z)$ is a boundary layer whose thickness, δ, goes to zero as $R \to \infty$. The bottom piece of $\tau(z)$ has a slope that is half the value of the

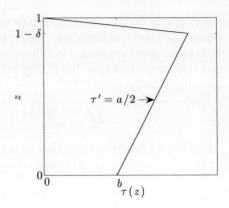

Fig. 2.3 Schematic of the class of background profiles, $\tau(z)$, that we consider. The parameters a, b, and δ are optimized, within some constraints, to maximize the lower bounds on the mean temperature

yet-unspecified constant a, a known trick [10, 27] for making the sign-indefinite term of $\mathscr{D}$ vanish outside the boundary layer.

For our three-parameter family of background profiles (2.45), the lower bound (2.40) becomes

$$\delta \langle T \rangle \geq b(2 - \delta) + \tfrac{a}{2}(1 - \delta) - \left(\tfrac{a^2}{4} + ab \right) \left(\tfrac{1}{\delta} - 1 \right) - \tfrac{b^2}{\delta}. \tag{2.46}$$

With a no-slip top in IH2 and any velocity conditions in IH1 or IH3, it can be shown that $\mathscr{D} \geq 0$ is satisfied when δ is no larger than [15, 27]

$$\delta^4 = \begin{cases} 64\,a\,R^{-1} & \text{IH1, IH3} \\ 32\,a\,R^{-1} & \text{IH2.} \end{cases} \tag{2.47}$$

We choose this δ because the tightest bounds result from choosing δ as large as possible.

In the IH2 and IH3 cases, we are free to choose the parameters δ, a, and b to maximize the lower bound (2.46) subject to (2.47). In IH1, the lower boundary condition requires that $b = 0$, so we are free to choose only δ and a. This maximization is carried out for IH1 and IH2 in [27] and [15], respectively, and the procedure for IH3 is analogous. The resulting optimal parameters are

$$\delta^* = \begin{cases} 4R^{-1/3} \\ 12^{1/3}R^{-1/3} \\ 2 \cdot 3^{1/3}R^{-1/3} \end{cases} \quad a^* = \begin{cases} 4R^{-1/3} \\ \tfrac{3 \cdot 12^{1/3}}{8}R^{-1/3} \\ \tfrac{3^{4/3}}{4}R^{-1/3} \end{cases} \quad b^* = \begin{cases} 0 & \text{IH1} \\ \tfrac{5 \cdot 12^{1/3}}{16}R^{-1/3} & \text{IH2} \\ \tfrac{5 \cdot 3^{1/3}}{8}R^{-1/3} & \text{IH3,} \end{cases} \tag{2.48}$$

for which the lower bound (2.46) becomes

$$\delta\langle T\rangle \geq \begin{cases} R^{-1/3} - & 4R^{-2/3} & \text{IH1} \\ \frac{9}{8}\left(\frac{3}{2}\right)^{1/3}R^{-1/3} - \frac{89}{64}\left(\frac{3}{2}\right)^{2/3}R^{-2/3} & \text{IH2} \\ \frac{3^{7/3}}{8}R^{-1/3} - & \frac{89\cdot 3^{2/3}}{64}R^{-2/3} & \text{IH3.} \end{cases} \qquad (2.49)$$

At large R, the leading-order terms of the bounds dominate:

$$\delta\langle T\rangle \gtrsim \begin{cases} R^{-1/3} & \text{IH1} \\ 1.28\,R^{-1/3} & \text{IH2} \\ 1.62\,R^{-1/3} & \text{IH3.} \end{cases} \qquad (2.50)$$

When Lu et al. [27] proved the above bound for IH1, they also raised the prefactor from 1 to 1.09 by generalizing the ansatz of $\tau(z)$ to include a bottom boundary layer, although this required solving an algebraic equation numerically. Their proof carries through for the IH3 case also, so they in fact proved the asymptotic lower bound of $1.09\,R^{-1/3}$ for both IH1 and IH3. By dropping the condition $\tau(0) = 0$ in IH3, where it is not needed, we have raised the prefactor to 1.62. Optimizing $\tau(z)$ beyond our limited ansatz would lower the prefactors of the bounds, but results of numerically optimizing $\tau(z)$ in the RB1 case suggest that the scaling of the bounds would not change [32].

2.3.2 Similarities Between RB and IH Bounds

A main virtue of the way we have defined the Nusselt numbers N and $\tilde{N}$ and the diagnostic Rayleigh number Ra and $\widetilde{Ra}$ is that bounds for the various configurations all have the same scaling when expressed using these quantities. Recalling that the definitions (1.33) of $\tilde{N}$ are inversely proportional to $\delta\langle T\rangle$, and that $\widetilde{Ra} = R/\tilde{N}$ in IH convection, we see that the asymptotic bounds (2.50) on $\delta\langle T\rangle$ become

$$\tilde{N} \lesssim \begin{cases} 0.025\,\widetilde{Ra}^{1/2} & \text{IH1} \\ 0.132\,\widetilde{Ra}^{1/2} & \text{IH2} \\ 0.094\,\widetilde{Ra}^{1/2} & \text{IH3.} \end{cases} \qquad (2.51)$$

These upper bounds on $\tilde{N}$ appear in Table 2.1 at the start of this chapter, along with the best known bounds on N in the RB configurations. The RB bounds have different prefactors but the same exponent, scaling proportionally to $Ra^{1/2}$. The RB1 prefactor in Table 2.1 comes from the improvement on [10] by Plasting and Kerswell [32], who also showed that their bound could not be improved without additional constraints. The prefactor in the other two RB cases comes from [29], where the analysis was aimed at RB2 but carries through for RB3 also.

Upper bounds with an exponent of $1/2$ are the best available for three-dimensional convection in general, but bounds with smaller exponents have been proven in special cases. Here too, analogies hold between various configurations if results are stated in terms of our diagnostic parameters. When the boundaries are free-slip, and either $Pr = \infty$ or the flow is two-dimensional, upper bounds with exponents of $5/12$ have been proven for RB1 [47, 48] and IH1 [46, 48]. When $Pr = \infty$ with no-slip boundaries, the best known bounds on N scale like $Ra^{1/3}(\log\log Ra)^{1/3}$ in RB1 [30] and like $Ra^{1/3}(\log Ra)^{1/2}$ in RB2 and RB3 [49], and the best known bound on $\tilde{N}$ scale like $\widetilde{Ra}^{1/3}(\log\widetilde{Ra})^{1/3}$ in IH1 [46]. Bounds with exponents smaller than $1/2$ are yet to be reported for the other RB or IH configurations.

Now that we have seen how the background method works, we can understand why it is challenging in the IH cases to prove upper bounds on $\langle wT \rangle$. This quantity is related to $\langle |\nabla \mathbf{u}|^2 \rangle$ by the velocity power integral (2.37) but is not is not related a priori to $\langle |\nabla T|^2 \rangle$ in IH convection. However, the parameter-dependent bounds that have been proven for convective models all amount to bounds on $\langle |\nabla T|^2 \rangle$ and rely on a background decomposition of the temperature field. Bounding $\langle |\nabla \mathbf{u}|^2 \rangle$ instead suggests a background decomposition of the velocity field, which has been carried out for shear flows (e.g., in [9]) but not for convection.

References

1. Ames, K.A., Straughan, B.: Penetrative convection in fluid layers with internal heat sources. Acta Mech. **85**, 137–148 (1990)
2. Busse, F.H.: On Howard's upper bound for heat transport by turbulent convection. J. Fluid Mech. **37**(3), 457–477 (1969)
3. Busse, F.H.: Remarks on the critical value $P_c = 0.25$ of the Prandtl number for internally heated convection found by Tveitereid and Palm. Eur. J. Mech. B/Fluids **47**, 32–34 (2014)
4. Chandrasekhar, S.: Hydrodynamic and Hydromagnetic Stability. Dover, New York (1981)
5. Chapman, C.J., Childress, S., Proctor, M.R.E.: Long wavelength thermal convection between non-conducting boundaries. Earth Planet. Sci. Lett. **51**, 362–369 (1980)
6. Chapman, C.J., Proctor, M.R.E.: Nonlinear Rayleigh–Bénard convection between poorly conducting boundaries. J. Fluid Mech. **101**(04), 759–782 (1980)
7. Chernyshenko, S.I., Goulart, P., Huang, D., Papachristodoulou, A.: Polynomial sum of squares in fluid dynamics: a review with a look ahead. Philos. Trans. R. Soc. A **372**, 20130350 (2014)
8. Childress, S., Spiegel, E.A.: Pattern formation in a suspension of swimming microorganisms: nonlinear aspects. In: Givoli, D., Grote, M.J., Papanicolaou, G.C. (eds.) A Celebr. Math. Model. Kluwer Academic Publishers, New York (2004)
9. Constantin, P., Doering, C.R.: Variational bounds on energy dissipation in incompressible flows: Shear flow. Phys. Rev. E **49**(5), 4087–4099 (1994)
10. Constantin, P., Doering, C.R.: Variational bounds on energy dissipation in incompressible flows. III. Convection. Phys. Rev. E **53**(6), 5957–5981 (1996)
11. Debler, W.R.: The onset of laminar natural convection in a fluid with homogenously distributed heat sources. Ph.D. thesis, University of Michigan (1959)
12. Doering, C.R., Constantin, P.: Energy dissipation in shear driven turbulence. Phys. Rev. Lett. **69**(11), 1648–1651 (1992)

13. Doering, C.R., Gibbon, J.D.: Applied Analysis of the Navier-Stokes Equations. Cambridge University Press, Cambridge (1995)
14. Galdi, G.P., Straughan, B.: Exchange of stabilities, symmetry, and nonlinear stability. Arch. Ration. Mech. Anal. **89**(3), 211–228 (1985)
15. Goluskin, D.: Internally heated convection beneath a poor conductor. J. Fluid Mech. **771**, 36–56 (2015)
16. Herron, I.H.: On the principle of exchange of stabilities in Rayleigh-Bénard convection, II - No-slip boundary conditions. Ann. dell'Università di Ferrara **IL**, 169–182 (2003)
17. Howard, L.N.: Heat transport by turbulent convection. J. Fluid Mech. **17**(3), 405–432 (1963)
18. Howard, L.N.: Bounds on flow quantities. Annu. Rev. Fluid Mech. **4**, 473–494 (1972)
19. Ishiwatari, M., Takehiro, S.I., Hayashi, Y.Y.: The effects of thermal conditions on the cell sizes of two-dimensional convection. J. Fluid Mech. **281**, 33–50 (1994)
20. Jeffreys, H.: Some cases of instability in fluid motion. Proc. R. Soc. A **118**, 195–208 (1928)
21. Joseph, D.D.: On the stability of the Boussinesq equations. Arch. Ration. Mech. Anal. **20**(1), 59–71 (1965)
22. Joseph, D.D.: Stability of Fluid Motions I-II. Springer, New York (1976)
23. Kaiser, R., Tilgner, A., Von Wahl, W.: A generalized energy functional for plane Couette flow. SIAM J. Math. Anal. **37**(2), 438–454 (2005)
24. Kulacki, F.A., Goldstein, R.J.: Hydrodynamic instability in fluid layers with uniform volumetric energy sources. Appl. Sci. Res. **31**(2), 81–109 (1975)
25. Kulacki, F.A., Nagle, M.E.: Natural convection in a horizontal fluid layer with volumetric energy sources. J. Heat Transfer **97**, 204–211 (1975)
26. Low, A.R.: On the criterion for stability of a layer of viscous fluid heated from below. Proc. R. Soc. A **125**(796), 180–195 (1929)
27. Lu, L., Doering, C.R., Busse, F.H.: Bounds on convection driven by internal heating. J. Math. Phys. **45**(7), 2967–2986 (2004)
28. McKenzie, D.P., Roberts, J.M., Weiss, N.O.: Convection in the earth's mantle: towards a numerical simulation. J. Fluid Mech. **62**(3), 465–538 (1974)
29. Otero, J., Wittenberg, R.W., Worthing, R.A., Doering, C.R.: Bounds on Rayleigh-Bénard convection with an imposed heat flux. J. Fluid Mech. **473**, 191–199 (2002)
30. Otto, F., Seis, C.: Rayleigh-Bénard convection: Improved bounds on the Nusselt number. J. Math. Phys. **52**(8), 083702 (2011)
31. Pellew, A., Southwell, R.V.: On maintained convective motion in a fluid heated from below. Proc. R. Soc. A **176**, 312–343 (1940)
32. Plasting, S.C., Kerswell, R.R.: Improved upper bound on the energy dissipation rate in plane Couette flow: the full solution to Busse's problem and the Constantin-Doering-Hopf problem with one-dimensional background field. J. Fluid Mech. **477**, 363–379 (2003)
33. Rayleigh, Lord: On convection currents in a horizontal layer of fluid, when the higher temperature is on the under side. Philos. Mag. **32**(192), 529–546 (1916)
34. Roberts, P.H.: Convection in horizontal layers with internal heat generation. Theory. J. Fluid Mech. **30**(01), 33–49 (1967)
35. Schwiderski, E.W.: Bifurcation of convection in internally heated fluid layers. Phys. Fluids **15**, 1882–1898 (1972)
36. Serrin, J.: On the stability of viscous fluid motions. Arch. Ration. Mech. Anal. **3**(1), 1–13 (1959)
37. Sparrow, E.M., Goldstein, R.J., Jonsson, V.K.: Thermal instability in a horizontal fluid layer: effect of boundary conditions and non-linear temperature profile. J. Fluid Mech. **18**(04), 513–528 (1964)
38. Straughan, B.: Continuous dependence on the heat source and non-linear stability for convection with internal heat generation. Math. Methods Appl. Sci. **13**, 373–383 (1990)
39. Straughan, B.: The Energy Method, Stability, and Nonlinear Convection, 2 edn. Springer, New York (2004)
40. Thirlby, R.: Convection in an internally heated layer. J. Fluid Mech. **44**(04), 673–693 (1970)
41. Trefethen, L.N.: Spectral Methods in MATLAB. SIAM, Philadelphia (2000)

42. Tveitereid, M.: Thermal convection in a horizontal fluid layer with internal heat sources. Int. J. Heat Mass Transfer **21**, 335–339 (1978)
43. Tveitereid, M., Palm, E.: Convection due to internal heat sources. J. Fluid Mech. **76**(03), 481 (1976)
44. Veronis, G.: Penetrative convection. Astrophys. J. **137**, 641–663 (1962)
45. Watson, P.M.: Classical cellular convection with a spatial heat source. J. Fluid Mech. **32**, 399 (1968)
46. Whitehead, J.P., Doering, C.R.: Internal heating driven convection at infinite Prandtl number. J. Math. Phys. **52**(9), 093101 (2011)
47. Whitehead, J.P., Doering, C.R.: Ultimate state of two-dimensional Rayleigh-Bénard convection between free-slip fixed-temperature boundaries. Phys. Rev. Lett. **106**(24), 244501 (2011)
48. Whitehead, J.P., Doering, C.R.: Rigid bounds on heat transport by a fluid between slippery boundaries. J. Fluid Mech. **707**, 241–259 (2012)
49. Whitehead, J.P., Wittenberg, R.W.: A rigorous bound on the vertical transport of heat in Rayleigh-Bénard convection at infinite Prandtl number with mixed thermal boundary conditions. J. Math. Phys. **55**(9), 093104 (2014)

Chapter 3
Internally Heated Convection Experiments and Simulations

Abstract Laboratory experiments and numerical simulations studying internally heated convection are reviewed. The emphasis is on quantitative results, especially integral quantities important to heat transport and their dependence on the Rayleigh number, which is proportional to the heating rate. For all experiments and three-dimensional simulations, the various measures of fluid temperature can be fit to powers of the rate of volumetric heating. The exponents of these fits range from 0.75 to 0.77 when the bottom is insulating, and they range from 0.78 to 0.82 when the top and bottom are fixed at equal temperatures. In the latter configuration, the fraction of internally produced heat flowing outward across the bottom boundary falls quite slowly as heating is strengthened. When this fraction is fit to a power of the heating rate, the fit exponents lie between -0.049 and -0.099.

The first two chapters have summarized features of heat transport in IH convection and RB convection that can be ascertained analytically from the Boussinesq equations. In this final chapter we summarize findings on IH convection from physical and computational experiments. Analogous results for RB convection are described only minimally, as the experimental literature on RB convection is vast and has been reviewed elsewhere (e.g., [1, 18, 26, 52, 68]).

Precise laboratory experiments on IH convections are inherently more difficult to carry out than similar experiments on RB convection. Both require maintaining the chosen thermal boundary conditions, but IH experiments also require producing heat internally in a controlled way. If our simple models are to apply, the heat production should be constant and uniform. In most experiments, the internal heating has been achieved by Joule heating, where the working fluid is an electrolytic solution that is heated by passing current through it. Two sets of experiments [50, 59] used a different method, wherein heating elements were distributed throughout the domain. Although neither method heats uniformly, it is possible that rapid mixing by strong convection limits the influence of non-uniformity. This is supported by the fairly good agreement between non-uniformly heated experiments and uniformly heated simulations.

Numerical simulations of IH convection avoid unknown variations in heating rate or material properties. However, the majority of numerical studies were carried out several decades ago and were limited to 2D and fairly small R. The larger values

© Springer International Publishing Switzerland 2016
D. Goluskin, *Internally Heated Convection and Rayleigh-Bénard Convection*,
SpringerBriefs in Applied Sciences and Technology,
DOI 10.1007/978-3-319-23941-5_3

of R accessible with modern computers have been simulated only a few times, and much of the parameter space that could now be reached has yet to be explored.

Experimental findings before 1985 are collected in the review of Kulacki and Richards [49], who discuss findings on IH1, IH3, and some similar configurations. The slightly later review of Cheung and Chawla [17] adds various scaling arguments for heat transport. Nourgaliev et al. [55] summarize heat fluxes in these same early experiments, as well as in experiments with curved geometries and cooled side walls.

A number of experiments have examined heat transport quantitatively, and a number of others have focused on qualitative pattern formation near the onset of convection. Here we cite studies of both types but focus on quantitative findings, and we restrict ourselves to experiments that closely resemble one of the IH configurations defined in Fig. 1.1. Convection with internal heating has been studied also with various complications that we do not confront, such as cooled side walls [3–6, 14, 21, 25, 34, 51, 55, 67, 71], non-uniform heating [45, 60, 73, 76], self-gravitating spheres [8, 9, 37, 61–63], and hybrid configurations driven both internally and by the boundary conditions [2, 7, 12, 19, 31, 35, 42–44, 54, 69, 70, 74, 83].

Section 3.1 addresses IH3, the last of the three IH configurations in Fig. 1.1. Section 3.2 addresses IH1, which is in some ways more complicated than IH3. We are not aware of any heat transport findings on the IH2 configuration, though 2D simulations have been carried out to study scale selection [33, 38]. Section 3.3 suggests directions for future work.

3.1   The IH3 Configuration

The internally heated configuration we call IH3, which is bounded above by a perfect conductor and below by a perfect insulator, has been the subject of numerous laboratory experiments [20, 24, 46, 48, 50, 58, 59, 66, 75, 77, 79], as well as computational studies both in 2D [22, 38, 54, 78] and in 3D [10, 11, 35, 36, 65, 78, 81]. Many of these investigations have focused on pattern formation and scale selection, which we do not discuss here. Our interest is in quantities relevant to heat transport, including the mean vertical temperature profile, $\overline{T}(z)$, and the mean temperature difference between the boundaries, $\delta\overline{T}$. No data are available on the mean fluid temperature, $\delta\langle T \rangle$.

Computational studies of IH3 convection that report $\overline{T}(z)$ or $\delta\overline{T}$ are all several decades old. Most are limited to the steady states that are stable at modest R [54, 65, 78, 81]. At larger R, unsteady 2D simulations have been carried out using a turbulence closure model [23] and by direct numerical simulation (DNS) [22], although some runs in the latter study seem under-resolved. As far as we know, unsteady IH3 convection has not been simulated in 3D. The largest R that has been reached in 2D DNS of IH3 [22] could be greatly exceeded in 3D DNS on modern parallel computers.

Laboratory experiments, most of which were carried out in the 1970s, furnish nearly everything we know about $\overline{T}(z)$ and $\delta\overline{T}$ in IH3 convection at large R [24, 46, 48, 50, 58, 59]. These findings are subject to the uncertainties inherent to IH experiments, so there is cause to repeat them numerically. The largest R reached in past laboratory experiments of IH3 could now be approached by 3D DNS, albeit in a smaller spatial domain.

3.1.1 Temperature Profiles

Several authors have reported mean vertical temperature profiles. It is simple to obtain $\overline{T}(z)$ in numerical studies by averaging steady flows horizontally [54, 78, 81] or averaging unsteady flows both horizontally and temporally [22, 23]. In laboratory experiments, vertical profiles have been obtained by measuring temperatures at fixed points and averaging only over time [58]. If the flow is horizontally isotropic in a statistical sense, and time averages are sufficiently long, then the same mean profile would be obtained whether or not horizontal averages are also taken. This is generally expected to be true at large R when side walls are absent or negligible. Some transient profiles have been reported also [46, 48], but these do not bear directly on the infinite-time averages we seek.

Figure 3.1 shows $\overline{T}(z)$ profiles for 2D steady states computed by Thirlby [78] for $Pr = 6.8$ and relatively small R. As R is raised and convection strengthens, the dimensionless temperature decreases, and the interior becomes closer to isothermal. When convection is sufficiently strong, the maximum value of $\overline{T}(z)$ occurs inside the layer, rather than at the bottom boundary. At still larger R, where convection is stronger and unsteady, the experimentally measured profiles of Ralph and Roberts [58] follow similar trends but are closer to isothermal in the interior, lacking the pronounced temperature inversion found in the steady states of Fig. 3.1. The

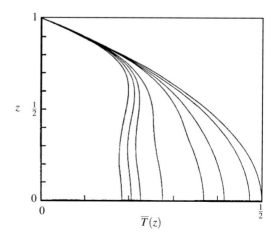

Fig. 3.1 Numerically computed mean temperature profiles, $\overline{T}(z)$, for steady 2D convection between no-slip boundaries. The Prandtl number is 6.8. The rightmost profile is that of the static state. The others, from *right* to *left*, are for $R = 10^3 \times (3, 4, 5, 10, 20, 30, 52)$. The figure is adapted from Fig. 2 of Thirlby [78]

unsteady motions responsible for homogenizing temperature outside the thermal boundary layer are evident in the IH3 temperature field of Fig. 1.2c: plumes emerge from the unstably stratified upper boundary layer and strongly mix fluid in the rest of the domain.

3.1.2 Mean Temperature Differences

The difference by which the dimensionless temperature at any point exceeds its average value at the top boundary, $T - \overline{T}_T$, tends to decrease as IH convection strengthens. This fact underlies the two related but distinct measures of convective strength discussed in Sect. 1.6: $\delta\langle T\rangle$, which is the average of $T - \overline{T}_T$ over time and the entire volume, and $\delta\overline{T}$, which is its average over time and the bottom boundary. (Recall that $\delta\overline{T}$ is also the mean vertical conduction, and in IH3 it is tied to the mean vertical convection, $\langle wT\rangle$, by the relation $\delta\overline{T} + \langle wT\rangle = 1/2$.) Both $\delta\overline{T}$ and $\delta\langle T\rangle$ are maximal in the static state and most likely approach zero in convective flows as $R \to \infty$.

A primary question for experimentalists is *how quickly* $\delta\overline{T}$ and $\delta\langle T\rangle$ fall as R is raised, along with how this answer is affected by the velocity boundary conditions, Prandtl number, and geometry. If $\delta\overline{T}$ and $\delta\langle T\rangle$ vary approximately as powers of R when other parameters are held constant, data can be captured by fits of the form

$$\delta\overline{T} \sim aR^{-\alpha} \qquad\qquad \delta\langle T\rangle \sim bR^{-\beta}. \qquad (3.1)$$

In the IH3 case, the Nusselt numbers and diagnostic Rayleigh numbers defined in Sects. 1.6.4 and 1.6.5 are

$$N = 1/2\delta\overline{T} \qquad\qquad Ra = R/N \qquad (3.2)$$

$$\tilde{N} = 1/3\delta\langle T\rangle \qquad\qquad \widetilde{Ra} = R/\tilde{N}. \qquad (3.3)$$

Various authors have considered quantities like N in past studies of IH3, and Fiedler and Wille [24] considered both N and Ra together. Restated in terms of the diagnostic variables, the fits of expression (3.1) become

$$N \sim cRa^{\gamma} \qquad\qquad \tilde{N} \sim d\widetilde{Ra}^{\delta}, \qquad (3.4)$$

were $\gamma = \alpha/(1-\alpha)$, $\delta = \beta/(1-\beta)$, $c = (2a)^{-1/(1-\alpha)}$, and $d = (3b)^{-1/(1-\beta)}$.

The mean temperature difference $\delta\overline{T}$ has been measured in a number of experiments. Table 3.1 summarizes past fits of the form $\delta\overline{T} \sim aR^{-\alpha}$, along with their corresponding re-expressions as fits of the form $N \sim cRa^{\gamma}$. Ranges of Pr and R are also given. The stated ranges of R are those over which the data have been fit. The Prandtl number would ideally be held constant as R is changed, but slight variations are unavoidable in the laboratory. Numerical studies do not suffer from

Table 3.1 Summary of IH3 experiments and simulations reporting approximate power-law dependence of $\delta\overline{T}$ on R

	Pr	R	$\delta\overline{T}$ fit	N fit
Laboratory experiments				
Fiedler and Wille [24]	6–7	10^4–10^7	$1.90\,R^{-0.228}$	$0.177\,Ra^{0.295}$
Ralph and Roberts [58]	6–7	$2.3\cdot10^5$–$6.0\cdot10^9$	$2.62\,R^{-0.25}$	$0.110\,Ra^{0.33}$
Kulacki and Nagle [48]	6.2–6.6	$1.5\cdot10^5$–$2.5\cdot10^9$	$3.28\,R^{-0.239}$	$0.0845\,Ra^{0.314}$
Kulacki and Emara [46]	2.7–6.9	$1.89\cdot10^3$–$2.17\cdot10^{12}$	$2.53\,R^{-0.227}$	$0.123\,Ra^{0.294}$
Ralph et al. [59]	6–7	10^9–$7\cdot10^9$	$a\,R^{-0.24}$	$c\,Ra^{0.32}$
Lee et al. [50]	0.71–0.74	$9.9\cdot10^9$–$3.3\cdot10^{11}$	$2.84\,R^{-0.247}$	$0.0996\,Ra^{0.328}$
Simulations (2D DNS)				
Mckenzie et al. [54] (free-slip, steady)	∞	$1.2\cdot10^4$–$7.0\cdot10^5$	$a\,R^{-0.26}$	$c\,Ra^{0.35}$
Emara and Kulacki [22] (free-slip top)	6.5	$5\cdot10^4$–$5\cdot10^8$	$1.07\,R^{-0.182}$	$0.397\,Ra^{0.222}$
Emara and Kulacki [22] (no-slip)	6.5	$5\cdot10^3$–$5\cdot10^8$	$2.38\,R^{-0.223}$	$0.134\,Ra^{0.287}$
Olwi [56] (no-slip, steady)	6.5	10^4–10^8	$3.07\,R^{-0.255}$	$0.0876\,Ra^{0.342}$

Internal heating was achieved by electric current in the first four experiments and by heating elements in the last two. The Prandtl number range 6–7 is an estimate for experiments that used aqueous solutions but did not report Pr measurements [24, 58, 59]

this uncertainty, but the simulation results in Table 3.1 nonetheless must be regarded with care since they all are 2D and seem to be somewhat under-resolved at larger R.

The decay rates of $\delta\overline{T}$ reported for the six laboratory experiments in Table 3.1 fall between $\alpha = 0.227$ and $\alpha = 0.25$. This means that the dimensional temperature difference between the boundaries, $\delta\overline{T}\Delta$, *grows* with the volumetric heating at rates between $H^{0.75}$ and $H^{0.773}$.

When the $\delta\overline{T}$ fits in Table 3.1 are restated in the form $N \sim c\,Ra^{\gamma}$, the exponents fall between $\gamma = 0.294$ and $\gamma = 0.33$. This range agrees very well with the analogous range of γ measured in RB1 experiments, where fits still take the form $N \sim c\,Ra^{\gamma}$, but with $N := 1 + \langle wT \rangle$ and $Ra := R$ (cf. Sect. 1.6.4). The RB1 exponents summarized in Table 1 of [29] lie between 0.25 and 0.33, excluding the very small Pr values for which corresponding IH3 data are unavailable. Exponents larger than 0.33 have sometimes been measured in RB1 experiments at very large R [13, 32], but no IH3 experiments have reached such R values. The similarity between measured values of γ in RB1 and IH3 is one of the analogies brought out by our chosen definitions of N and Ra.

No data have been reported on the volume-averaged quantity $\delta\langle T \rangle$, so we cannot say exactly what exponents would emerge from fits of the form $\delta\langle T \rangle \sim b\,R^{-\beta}$ or $\widetilde{N} \sim d\,\widetilde{Ra}^{\delta}$. We can reasonably estimate the exponents, however, since the temperature profiles that have been reported are close to isothermal outside their boundary layers. This suggests that the values of $\delta\langle T \rangle$ and $\delta\overline{T}$ become ever closer as $R \to \infty$, in which case $\alpha \approx \beta$ and $\gamma \approx \delta$ for sufficiently large R. This speculation remains to be tested since volume averages like $\delta\langle T\rangle$ are difficult to measure in the laboratory. They are easy to extract from simulations, however, and we hope that future numerical studies will report $\delta\langle T\rangle$.

Whereas we have data on $\delta\overline{T}$ but not on $\delta\langle T\rangle$—or, equivalently, on N but not on $\tilde{N}$—the state of affairs for analytical bounds is just the opposite. We have conjectured in Chap. 1, but have not proven, than N obeys an upper bound of the form $c\,Ra^{1/2}$. The experimental exponents β in Table 3.1 are all smaller than 1/2 and thus consistent with this conjecture. On the other hand, we *have* proven in Chap. 2 that $\tilde{N}$ can grow no faster than $0.093\,\widetilde{Ra}^{1/2}$, but no data on $\delta\langle T\rangle$ have been reported for the IH3 configuration.

3.2 The IH1 Configuration

The internally heated configuration we call IH1, which is bounded above and below by perfect conductors of equal temperature, has been studied in the laboratory [39–41, 47, 50, 53, 59], as well as numerically both in 2D [22, 27, 53, 57, 74, 80] and in 3D [28, 30, 84]. Almost all of these studies have reported quantitatively on heat transport in some way.

Numerical computations of IH1 include both steady states [57, 74, 80] and DNS. Whereas DNS of the IH3 configuration has been limited to a single 2D study, DNS of the IH1 configuration has been carried out up to fairly large R in both 2D [22, 27] and 3D [28, 30, 84].

3.2.1 Temperature Profiles

Mean vertical temperature profiles have been reported in a number of studies. Numerical studies provide profiles, $\overline{T}(z)$, that are averaged horizontally and, if the simulations are unsteady, over time as well [22, 27, 28, 30, 53, 57, 74, 84]. In the laboratory, profiles measured pointwise by temperature probes are averaged only over time [50, 59], while profiles gleaned from interferograms are instantaneous but effectively averaged over a horizontal direction [47, 53].

Figure 3.2a shows $\overline{T}(z)$ profiles from the 3D DNS data of Goluskin and van der Poel [28]. As in the IH3 configuration, raising R strengthens convection, which decreases the dimensionless temperature and brings the interior closer to isothermality. When R is large enough for thermal boundary layers to be discernible, the top boundary layer is visibly thinner than the bottom one. This reflects the up-down asymmetry of heat fluxes; more of the produced heat flows outward across the top boundary than across the bottom one, as quantified in the next subsection. The same basic features are evident in Fig. 3.2b, which shows an interferogram

Fig. 3.2 (**a**) Mean
temperature profiles, $\overline{T}(z)$,
from the 3D DNS of Goluskin
and van der Poel [28] for a
fluid with $Pr = 1$ between
no-slip boundaries. The
rightmost profile is that of the
static state. The others, from
right to *left*, are for $R = 10^6$,
10^7, 10^8, 10^9, and 10^{10}. (**b**)
An interferogram from the
experiments of Kulacki and
Goldstein [47] with
$R = 1.5 \cdot 10^5$ and $Pr = 5.8$.
Any curve of constant color
acts approximately as a graph
of instantaneous temperature
averaged over one horizontal
direction

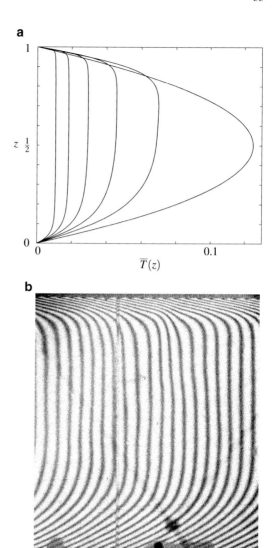

from the experiments of Kulacki and Goldstein [47]. The interferogram measures
horizontally averaged optical properties of the fluid that vary with its temperature,
and a line of constant color can be interpreted as a temperature profile.

The IH1 configuration stands out from the other RB and IH models we have discussed in that there is a stably stratified thermal boundary layer. The configuration
thus provides a simple instance of *penetrative convection*, wherein buoyancy forces
in an unstably stratified region drive motions that penetrate into a stably stratified
region. The temperature field of Fig. 1.2b reflects the dissimilarity between the
unstably stratified upper boundary layer and the stably stratified lower one. Mixing
of the cold upper layer with the warmer interior is accomplished by buoyantly

driven cold plumes. At large R, the cold *lower* layer also can mix with the warmer interior. This mixing is driven by shear forces, rather than by buoyancy, and it occurs when the interior turbulence pulls cold eddies off the bottom boundary layer.

3.2.2 Maximum Temperatures, Mean Temperatures, and Asymmetry

The mean fluid temperature, $\delta\langle T \rangle$, behaves in IH1 convection much like it does in IH3 convection, assuming its maximum value in the static state and falling as R is raised. On the other hand, the mean temperature change between the boundaries, $\delta\overline{T}$, differs completely between the two configurations. Whereas in IH3 $\delta\overline{T}$ behaves rather like $\delta\langle T \rangle$, in IH1 it is identically zero. The role that $\delta\overline{T}$ plays in IH3 is instead approximated in IH1 by $\overline{T}_{\max}$, the maximum value that $\overline{T}(z)$ assumes over the layer. Whereas $\delta\overline{T}$ equals the mean upward conduction across the entire layer, $\overline{T}_{\max}$ captures the mean *outward* conduction, as described in Sect. 1.6.4.1. The quantity $\overline{T}_{\max}$ has been reported in many studies of IH1 since it is easier to estimate in the laboratory than $\delta\langle T \rangle$. However, $\overline{T}_{\max}$ does not arise as easily as $\delta\langle T \rangle$ in analytical expressions.

In addition to $\delta\langle T \rangle$ and $\overline{T}_{\max}$, IH1 convection is naturally characterized by the extent to which the flow creates asymmetry between upward and downward heat fluxes. This asymmetry can be simply conveyed by the mean fractions of produced heat leaving across the top or bottom boundaries—$\mathscr{F}_T$ or $\mathscr{F}_B$. As described in Sect. 1.6.2, these fractions are related to the dimensionless convective flux, $\langle wT \rangle$, by

$$\mathscr{F}_T = \tfrac{1}{2} + \langle wT \rangle \qquad\qquad \mathscr{F}_B = \tfrac{1}{2} - \langle wT \rangle. \tag{3.5}$$

One can equivalently speak in terms of $\langle wT \rangle$, $\mathscr{F}_T$, or $\mathscr{F}_B$. Here we focus on $\mathscr{F}_B$ because it comes the closest to having a power-law dependence on R as in the regimes studied.

Despite their simplicity and analytical attractiveness, neither $\delta\langle T \rangle$ nor $\mathscr{F}_B$ has received much attention, although both quantities have been mentioned. Most authors have instead spoken in terms of top and bottom Nusselt numbers, here called N_T and N_B. The most common definitions of these numbers are

$$N_T := \frac{\mathscr{F}_T}{\overline{T}_{\max}} \qquad\qquad N_B := \frac{\mathscr{F}_B}{\overline{T}_{\max}}. \tag{3.6}$$

The above expressions are not normalized to be unity in the static state; instead, both are equal to 4. Data on N_T and N_B are typically fit to powers of R.

To keep measures of asymmetry and temperature as separate as possible, we prefer not to examine N_T and N_B. Instead, we use $\mathscr{F}_B$ as a measure of asymmetry and use $\overline{T}_{\max}$ and $\delta\langle T \rangle$—or their inverses, N and $\tilde{N}$—as measures of temperature. One undesirable feature of N_B is that it can initially drop below its static value as R

is raised since $\mathscr{F}_B$ can initially fall faster than $\overline{T}_{\max}$. For instance, this occurs in the data of Kulacki and Goldstein [47]. Such behavior prevents N_B from being well fit by a power of R near onset, and it is unlike the behavior of the RB Nusselt number, which cannot be smaller than its static value. Another disadvantage of using N_T and N_B is that their R-dependence differs only in regimes where $\mathscr{F}_B$ is changing significantly. If the decay of $\mathscr{F}_B$ stops, as in the 2D simulations of Goluskin and Spiegel [27], then N_T and N_B will both be dominated by the scaling of $1/\overline{T}_{\max}$, and slight changes in the asymmetry will not be captured well.

We would like to summarize past data on R-dependence with fits of the form

$$\overline{T}_{\max} \sim aR^{-\alpha} \qquad \delta\langle T \rangle \sim bR^{-\beta} \qquad \mathscr{F}_B \sim eR^{-\varepsilon}. \qquad (3.7)$$

Fits of the above form have been reported for all three quantities in [28] and for $\delta\langle T \rangle$ in [27]. In two other studies where the original data are available to us [47, 84], we have calculated fits to $\overline{T}_{\max}$ and $\mathscr{F}_B$. For the remaining studies, only fits to N_T and N_B are available. In these cases, we use the relations

$$\overline{T}_{\max} = \frac{1}{N_T + N_B} \qquad \mathscr{F}_B = \frac{N_B}{N_T + N_B}. \qquad (3.8)$$

The reported power-law fits to N_T and N_B define curves for $\overline{T}_{\max}$ and $\mathscr{F}_B$ that are not pure powers of R, so we have re-fit pure power laws to the latter curves.

Table 3.2 summarizes power-law fits to the R-dependence of $\overline{T}_{\max}$, $\delta\langle T \rangle$, and $\mathscr{F}_B$. The fits to $\overline{T}_{\max}$ are also stated in terms of N and Ra, and the fits to $\delta\langle T \rangle$ are also stated in terms of $\tilde{N}$ and $\widetilde{Ra}$. For the IH1 configuration, we have defined these diagnostic quantities in Sects. 1.6.4 and 1.6.5 as

$$N = 1/8\,\overline{T}_{\max} \qquad\qquad Ra = R/N \qquad (3.9)$$

$$\tilde{N} = 1/12\,\delta\langle T \rangle \qquad\qquad \widetilde{Ra} = R/\tilde{N}. \qquad (3.10)$$

The fits (3.7) to $\overline{T}_{\max}$ and $\delta\langle T \rangle$ imply fits to and N and $\tilde{N}$ of the form (3.4), where $\gamma = \alpha/(1-\alpha)$, $\delta = \beta/(1-\beta)$, $c = (8a)^{-1/(1-\alpha)}$, and $d = (12b)^{-1/(1-\beta)}$.

3.2.2.1   Maximum Temperatures

Fits of the form $\overline{T}_{\max} \sim aR^{-\alpha}$ are shown in Table 3.2. In all laboratory experiments and all simulations with no-slip boundaries, the exponent α lies between 0.180 and 0.224. This means that the dimensional maximum temperature, $\overline{T}_{\max}\Delta$, grows with the volumetric heating at rates between $H^{0.776}$ and $H^{0.820}$. In the sole study for which both $\overline{T}_{\max}$ and $\delta\langle T \rangle$ are reported [28], the decay of $\overline{T}_{\max}$ is slightly faster than the decay of $\delta\langle T \rangle$, the fit exponents being $\alpha = 0.217$ and $\beta = 0.204$, respectively. This makes sense since $\overline{T}_{\max}$ initially must "catch up" to $\delta\langle T \rangle$ as the temperature profile flattens. When the $\overline{T}_{\max}$ fits are restated in the form $N \sim cRa^\gamma$, the exponents

Table 3.2 Summary of IH1 experiments and simulations reporting approximate power-law dependence of $\overline{T}_{max}$, $\delta\overline{T}$, or $\mathscr{F}_B$ on R

	Pr	R	$\overline{T}_{max}$ fit	N fit	$\delta\langle T\rangle$ fit	$\tilde{N}$ fit	$\mathscr{F}_B$ fit
Laboratory experiments							
Kulacki and Goldstein [47]	5.7–6.3	$R_L - 2.4\cdot10^7$	$1.71R^{-0.180}$	$0.0958Ra^{0.219}$			$1.21R^{-0.0848}$
Jahn and Reineke [39, 53]	≈ 7	10^5-10^9	$1.96R^{-0.194}$	$0.0778Ra^{0.240}$			$1.36R^{-0.0988}$
Ralph et al. [59]	6–7	$3.7\cdot10^8 - 1.1\cdot10^{12}$	$5.39R^{-0.224}$	$0.0191Ra^{0.289}$			$0.692R^{-0.0494}$
Lee et al. [50]	0.71–0.74	$1.1\cdot10^{10} - 3.7\cdot10^{11}$	$3.86R^{-0.209}$	$0.0315Ra^{0.264}$			$2.48R^{-0.0947}$
Simulations (3D DNS)							
Wörner et al. [84]	7	10^5-10^8	$1.86R^{-0.186}$	$0.0847Ra^{0.229}$			$1.16R^{-0.0845}$
Goluskin and van der Poel [28]	1	$5\cdot10^7-2\cdot10^{10}$	$1.62R^{-0.217}$	$0.0379Ra^{0.277}$	$1.11R^{-0.204}$	$0.0386\widetilde{Ra}^{0.256}$	$0.803R^{-0.0554}$
Simulations (2D DNS)							
Jahn and Reineke [39, 53]	7	$1\cdot10^5-1\cdot10^9$	$2.20R^{-0.192}$	$0.0678Ra^{0.238}$			$1.19R^{-0.0854}$
Peckover and Hutchinson [57] (free-slip, steady)	8	$5.1\cdot10^4-1.4\cdot10^6$	$0.575R^{-0.104}$	$0.182Ra^{0.116}$			$0.953R^{-0.0752}$
Straus [74] (free-slip, steady)	∞	$10^5-3\cdot10^5$	$1.82R^{-0.217}$	$0.0795Ra^{0.277}$			$1.23R^{-0.100}$
Emara and Kulacki [22]	6.5	$5\cdot10^4-5\cdot10^7$	$1.96R^{-0.186}$	$0.0795Ra^{0.229}$			$1.02R^{-0.0672}$
Goluskin and Spiegel [27]	1	$10^8-2\cdot10^{10}$			$1.13R^{-0.200}$	$0.0384\widetilde{Ra}^{0.250}$	

Internal heating was achieved by heating elements in one laboratory experiment [50] and by electric current in the others. Fits to $\overline{T}_{max}$ and $\mathscr{F}_B$ are computed directly from data for a few studies [27, 28, 47, 84], while for other studies we have computed them from reported fits to N_T and N_B (see text). Simulations employ no-slip boundary conditions, except when specified otherwise range from $\gamma = 0.220$ to $\gamma = 0.289$. The bottom end of this range is smaller than any exponents found for the ordinary RB1 Rayleigh number, except at very small Pr [29].

3.2.2.2 Mean Temperatures

The quantity $\delta\langle T \rangle$ has been reported only in two numerical studies, and each gives a fit of the form $\delta\langle T \rangle \sim bR^{-\beta}$ for $Pr = 1$. Despite one study being 3D and the other 2D, the growth rates of $\delta\langle T \rangle$ with R are very similar, having exponents of $\beta = 0.204$ in 3D [28] and $\beta = 0.200$ in 2D [27]. This is reminiscent of the Nusselt number in RB convection, which is not much affected by dimensionality unless Pr is small [64, 82].

The dimensional mean temperature, $\delta\langle T \rangle\Delta$, grows with the volumetric heating proportionally to $H^{0.796}$ in 3D and to $H^{0.800}$ in 2D. When the $\delta\langle T \rangle$ fits are restated in the form $\widetilde{N} \sim d\widetilde{Ra}^{\delta}$, the exponents are $\delta = 0.256$ in 3D and $\delta = 0.250$ in 2D. These δ values are within the range of Nusselt number growth rates seen in RB1 convection, though they are at the lower end of that range (cf. Sect. 3.1.2). We cannot yet draw comparison with IH3 convection, for which no data on $\delta\langle T \rangle$ have been reported.

3.2.2.3 Asymmetry

The asymmetry between upward and downward heat fluxes in IH1 convection, as quantified by the fraction of heat that flows downward, $\mathscr{F}_B$, seems to have no analogues in our other five IH or RB configurations. First, this fraction changes with R much more slowly than any other integral quantity we have discussed. Second, the R-dependence of $\mathscr{F}_B$ can differ greatly between 2D and 3D, even when Pr is not small. This is because shear, rather than buoyancy, is the mechanism responsible for mixing the cooler lower boundary layer with the warmer interior. The asymmetry is generally greater in 3D than in 2D because the shear-driven mixing, which helps heat escape across the bottom boundary, is less effective in 3D [28].

A particularly simply question without an obvious answer is: as $R \to \infty$, what is the limit of $\mathscr{F}_B$? The extreme possibilities of either 0 or 1/2 seem most likely, although intermediate values are also plausible. In the highest-R simulation data available in 3D, $\mathscr{F}_B$ falls monotonically as R is raised [28]. The highest-R data available in 2D are quite different, except perhaps at large Pr [27, 28]. For instance, in the 2D simulations of Goluskin and Spiegel [27] with $Pr = 1$, the fraction $\mathscr{F}_B$ reaches a minimum of 0.33 near $R = 10^9$ and then increases as R is raised further. This non-monotonic R-dependence in 2D is yet another way that $\mathscr{F}_B$ stands apart from other quantities we have considered.

When $\mathscr{F}_B$ decreases monotonically as R is raised, as in all past 3D studies and some 2D ones, we can seek fits of the form $\mathscr{F}_B \sim eR^{-\varepsilon}$. Table 3.2 summarizes these fits, all of whose decay rates are quite small. The decay rates range from $\varepsilon = 0.0494$

to $\varepsilon = 0.0988$. It remains a mystery whether such decay will continue or reverse at larger R.

The dependence of $\mathscr{F}_B$ on Pr has been examined in two studies [27, 28]. The value of $\mathscr{F}_B$ seems to fall monotonically as Pr is raised, meaning that the asymmetry increases, until saturating at large Pr. The effect of Pr on the asymmetry is fairly strong—stronger than its effect on $\delta\langle T\rangle$ or $\overline{T}_{\max}$.

3.2.3   Scaling Arguments

Several scaling arguments have been put forth to explain the parameter-dependence of mean temperatures in IH convection for both the IH1 and IH3 configurations [15–17, 27]. In RB convection, the Nusselt number displays a wide diversity of scaling behavior in different regions of parameter space [72]. It is likely that the same is true of mean temperatures in IH convection since (inverses of) these temperatures have many parallels to the RB Nusselt number. This remains to be confirmed by a wider exploration of parameter space. If such a diversity of scaling can indeed be found in IH convection, then any broadly applicable scaling arguments must reflect this. For the standard RB1 configuration, the only arguments that attempt to capture the full range of scaling behavior are those put forth by Grossman and Lohse in [29] and subsequent papers (see [72]). The arguments of [29] carry through analogously for IH convection [27]. When the predicted scalings are phrased in terms of N and Ra, or $\widetilde{N}$ and $\widetilde{Ra}$, they are the same as the scalings predicted for the Nusselt number in the RB1 case. However, further work on scaling arguments is perhaps premature until data are available across a wider swath of parameter space.

3.3   Future Directions

In the future study of IH convection, the main task accessible to mathematical analysis is proving parameter-dependent bounds on key integral quantities. The only results of this kind are the R-dependent lower bounds on volume-averaged temperatures described in Sect. 2.3. We have conjectured in Sect. 1.6.3 that there should also exist R-dependent upper bounds on the mean convective flux, $\langle wT\rangle$. These would amount to lower bounds on the fraction of heat flowing downward in IH1 and on the mean temperature difference between the boundaries in IH2 and IH3. Bounds are lacking also for the maximum horizontally averaged temperature, $\overline{T}_{\max}$, that has often been measured in IH1 experiments. Bounds depending analytically on the Prandtl number are highly desirable as well.

There is much fertile ground for physical and computational experiments on IH convection. This is especially true for computation since most prior results are several decades old, so modern computers would be able to probe unexplored parameter regimes with relative ease. Neither the IH2 nor IH3 configuration has been simulated in 3D, and the DNS carried out in 2D has not approached the large R that are now computationally accessible. The IH1 configuration has been the subject of two DNS studies in 3D [28, 84], but a much wider exploration of parameter space is called for. The asymmetry between upward and downward heat fluxes in IH1 is particularly hard to predict; even its value as R approaches infinity is not certain. In each of the three IH configurations, simulating a wide range of R and Pr would produce a more global picture of how key integral quantities depend on the control parameters. The complicated parameter-dependence of Nusselt numbers in RB convection [1, 72] suggests that fitting integral quantities to pure powers of R will not suffice.

A combination of mathematical analysis, simulation, and physical experimentation will lead to a better understanding of the three internally heated configurations we have studied in this SpringerBrief. We hope that this, in turn, will lead to a better understanding of more complicated occurrences of IH convection. The many past studies of RB convection should prove useful in guiding future studies of IH convection, and to this end we have described a number of analogies between the two classes of flows. Still, the analogies are not perfect, and some consequences of internal heating cannot be foreseen. Judging by the complexity of RB convection, we expect that these novel aspects of IH convection will remain rich areas of inquiry for many more years.

References

1. Ahlers, G., Grossmann, S., Lohse, D.: Heat transfer and large scale dynamics in turbulent Rayleigh-Bénard convection. Rev. Mod. Phys. **81**(2), 503–537 (2009)
2. Ames, K.A., Straughan, B.: Penetrative convection in fluid layers with internal heat sources. Acta Mech. **85**, 137–148 (1990)
3. Arcidiacono, S., Ciofalo, M.: Low-Prandtl number natural convection in volumetrically heated rectangular enclosures III. Shallow cavity, AR=0.25. Int. J. Heat Mass Transf. **44**, 3053–3065 (2001)
4. Arcidiacono, S., Di Piazza, I., Ciofalo, M.: Low-Prandtl number natural convection in volumetrically heated rectangular enclosures II. Square cavity, AR=1. Int. J. Heat Mass Transf. **44**, 537–550 (2001)
5. Asfia, F.J., Dhir, V.K.: An experimental study of natural convection in a volumetrically heated spherical pool bounded on top with a rigid wall. Nucl. Eng. Des. **163**(3), 333–348 (1996)
6. Bergholz, R.F.: Natural convection of a heat generating fluid in a closed cavity. J. Heat Transfer **102**, 242–247 (1980)
7. Berlengiero, M., Emanuel, K.A., von Hardenberg, J., Provenzale, A., Spiegel, E.A.: Internally cooled convection: A fillip for Philip. Commun. Nonlinear Sci. Numer. Simul. **17**(5), 1998–2007 (2012)
8. Busse, F.H.: Patterns of convection in spherical shells. J. Fluid Mech. **72**(1), 67–85 (1975)
9. Busse, F.H., Riahi, N.: Patterns of convection in spherical shells. Part 2. J. Fluid Mech. **123**, 282–301 (1975)
10. Cartland Glover, G.M., Generalis, S.C.: Pattern competition in homogeneously heated fluid layers. Eng. Appl. Comput. Fluid Mech. **3**(2), 164–174 (2009)
11. Cartland Glover, G., Fujimura, K., Generalis, S.: Pattern formation in volumetrically heated fluids. Chaotic Model. Simul. **1**, 19–30 (2013)

12. Chapman, C.J., Childress, S., Proctor, M.R.E.: Long wavelength thermal convection between non-conducting boundaries. Earth Planet. Sci. Lett. **51**, 362–369 (1980)
13. Chavanne, X., Chilla, F., Castaing, B., Hebral, B., Chabaud, B., Chaussy, J.: Observation of the ultimate regime in Rayleigh-Bénard convection. Phys. Rev. Lett. **79**(19), 3648–3651 (1997)
14. Chen, S., Krafczyk, M.: Entropy generation in turbulent natural convection due to internal heat generation. Int. J. Therm. Sci. **48**(10), 1978–1987 (2009)
15. Cheung, F.B.: Natural convection in a volumetrically heated fluid layer at high Rayleigh numbers. Int. J. Heat Mass Transf. **20**(5), 499–506 (1977)
16. Cheung, F.B.: Heat source-driven thermal convection at arbitrary Prandtl number. J. Fluid Mech. **97**(4), 743–758 (1980)
17. Cheung, F.B., Chawla, T.C.: Complex heat transfer processes in heat-generating horizontal fluid layers. In: Annual review of numerical fluid mechanics and heat transfer, vol. 1, pp. 403–448. Hemisphere, New York (1987)
18. Chillà, F., Schumacher, J.: New perspectives in turbulent Rayleigh-Bénard convection. Eur. Phys. J. E **35**(7), 1–25 (2012)
19. Clever, R.M.: Heat transfer and stability properties of convection rolls in an internally heated fluid layer. J. Appl. Math. Phys. **28**, 585–597 (1977)
20. De la Cruz Reyna, S.: Asymmetric convection in the upper mantle. Geofis. Int. **10**, 49–56 (1970)
21. Di Piazza, I., Ciofalo, M.: Low-Prandtl number natural convection in volumetrically heated rectangular enclosures I. Slender cavity, AR=4. Int. J. Heat Mass Transf. **43**, 3027–3051 (2000)
22. Emara, A.A., Kulacki, F.A.: A numerical investigation of thermal convection in a heat-generating fluid layer. J. Heat Transf. **102**, 531–537 (1980)
23. Farouk, B.: Turbulent thermal convection in an enclosure with internal heat generation. J. Heat Transf. **110**(1), 126–132 (1988)
24. Fiedler, H.E., Wille, R.: Turbulente freie konvektion in einer horizontalen flüssigkeitsschicht mit volumen-wärmequelle. In: Proceeding of 4th International Heat Transfer Conference (1970)
25. Filippov, A.S.: Numerical simulation of experiments on turbulent natural convection of heat generating liquid in cylindrical pool. J. Eng. Thermophys. **20**(1), 64–76 (2011)
26. Getling, A.V.: Rayleigh-Bénard convection: structures and dynamics. World Scientific Publishing Co (1998)
27. Goluskin, D., Spiegel, E.A.: Convection driven by internal heating. Phys. Lett. A **377**(1-2), 83–92 (2012)
28. Goluskin, D., van der Poel, E.P.: Penetrative internally heated convection in two and three dimensions. Submitted. (2015)
29. Grossmann, S., Lohse, D.: Scaling in thermal convection: a unifying theory. J. Fluid Mech. **407**, 27–56 (2000)
30. Grötzbach, G.: Turbulent heat transfer in an internally heated fluid layer. In: The Third International Symposium on Refined Flow Modelling and Turbulence Measurements, vol. 2, p. 8. Tokyo (1988)
31. Hartlep, T., Busse, F.H.: Convection in an internally cooled fluid layer heated from below. Technical Representation, Center for Turnulence Research (2006)
32. He, X., Funfschilling, D., Nobach, H., Bodenschatz, E., Ahlers, G.: Transition to the ultimate state of turbulent Rayleigh-Bénard convection. Phys. Rev. Lett. **108**, 024502 (2012)
33. Hewitt, J.M., McKenzie, D.P., Weiss, N.O.: Large aspect ratio cells in two-dimensional thermal convection. Earth Planet. Sci. Lett. **51**, 370–380 (1980)
34. Horvat, A., Kljenak, I., Marn, J.: Two-dimensional large-eddy simulation of turbulent natural convection due to internal heat generation. Int. J. Heat Mass Transf. **44**(21), 3985–3995 (2001)
35. Houseman, G.: The dependence of convection planform on mode of heating. Nature **332**, 346–349 (1988)
36. Ichikawa, H., Kurita, K., Yamagishi, Y., Yanagisawa, T.: Cell pattern of thermal convection induced by internal heating. Phys. Fluids **18**(3), 038101 (2006)

37. Ingersoll, A.P., Porco, C.C.: Solar heating and internal heat flow on Jupiter. Icarus **35**, 27–43 (1978)
38. Ishiwatari, M., Takehiro, S.I., Hayashi, Y.Y.: The effects of thermal conditions on the cell sizes of two-dimensional convection. J. Fluid Mech. **281**, 33–50 (1994)
39. Jahn, M., Reineke, H.H.: Free convection heat transfer with internal heat sources, calculations and measurements. In: Proceedings of 5th International Heat Transfer Conference, pp. 74–78. Tokyo (1974)
40. Jaupart, C., Brandeis, G.: The stagnant bottom layer of convecting magma chambers. Earth Planet. Sci. Lett. **80**, 183–199 (1986)
41. Jaupart, C., Brandeis, G., Allègre, C.J.: Stagnant layers at the bottom of convecting magma chambers. Nature **308**, 535–538 (1984)
42. Joseph, D.D.: Subcritical Instability and Exchange of Stability in a Horizontal Fluid Layer. Phys. Fluids **11**(1968), 903–904 (1968)
43. Joseph, D.D., Shir, C.C.: Subcritical convective instability: Part 1. Fluid layers. J. Fluid Mech. **26**(4), 753–768 (1966)
44. Kolmychkov, V.V., Mazhorova, O.S., Shcheritsa, O.V.: Numerical study of convection near the stability threshold in a square box with internal heat generation. Phys. Lett. A **377**, 2111–2117 (2013)
45. Kondratenko, P.S., Nikolski, D.V., Strizhov, V.F.: Free-convective heat transfer in fluids with non-uniform volumetric heat generation. Int. J. Heat Mass Transf. **51**(7-8), 1590–1595 (2008)
46. Kulacki, F.A., Emara, A.A.: Steady and transient thermal convection in a fluid layer with uniform volumetric energy sources. J. Fluid Mech. **83**(2), 375–395 (1977)
47. Kulacki, F.A., Goldstein, R.J.: Thermal convection in a horizontal fluid layer with uniform volumetric energy sources. J. Fluid Mech. **55**(02), 271–287 (1972)
48. Kulacki, F.A., Nagle, M.E.: Natural convection in a horizontal fluid layer with volumetric energy sources. J. Heat Transf. **97**, 204–211 (1975)
49. Kulacki, F.A., Richards, D.E.: Natural convection in plane layers and cavities with volumetric energy sources. In: Natural Convection: Fundamentals and Applications, pp. 179–254. Hemisphere, New York (1985)
50. Lee, S.D., Lee, J.K., Suh, K.Y.: Boundary condition dependent natural convection in a rectangular pool with internal heat sources. J. Heat Transf. **129**(5), 679–682 (2007)
51. Liu, H., Zou, C., Shi, B., Tian, Z., Zhang, L., Zheng, C.: Thermal lattice-BGK model based on large-eddy simulation of turbulent natural convection due to internal heat generation. Int. J. Heat Mass Transf. **49**, 4672–4680 (2006)
52. Lohse, D., Xia, K.Q.: Small-scale properties of turbulent Rayleigh-Bénard convection. Annu. Rev. Fluid Mech. **42**(1), 335–364 (2010)
53. Mayinger, F., Jahn, M., Reineke, H.H., Steinberner, U.: Examination of thermohydraulic processes and heat transfer in a core melt. Technical Representation, Hannover Technical University, Hannover, Germany (1975)
54. McKenzie, D.P., Roberts, J.M., Weiss, N.O.: Convection in the earth's mantle: towards a numerical simulation. J. Fluid Mech. **62**(3), 465–538 (1974)
55. Nourgaliev, R.R., Dinh, T.N., Sehgal, B.R.: Effect of fluid Prandtl number on heat transfer characteristics in internally heated liquid pools with Rayleigh numbers up to 10^{12}. Nucl. Eng. Des. **169**, 165–184 (1997)
56. Olwi, I.A., Kulacki, F.A.: Numerical simulation of the transient convection process in a volumetrically heated fluid layer. In: Proceeding of ASME, p. 185 (1995)
57. Peckover, R.S., Hutchinson, I.H.: Convective rolls driven by internal heat sources. Phys. Fluids **17**(7), 1369–1371 (1974)
58. Ralph, J.C., Roberts, D.N.: Free convection heat transfer measurements in horizontal liquid layers with internal heat generation. Technical Representation, UKAEA (1974)
59. Ralph, J.C., McGreevy, R., Peckover, R.S.: Experiments in tubulent thermal convection driven by internal heat sources. In: Spalding, D.B., Afgan, N. (eds.) Heat Transfer and Turbulent Buoyant Convection: Studies and Applications for Natural Environment, Buildings, Engineering Systems, pp. 587–599. Hemisphere, New York (1977)

60. Riahi, N.: Nonlinear convection in a horizontal layer with an internal heat source. J. Phys. Soc. Japan **53**(12), 4169–4178 (1984)
61. Riahi, D.N., Busse, F.H.: Pattern generation by convection in spherical-shells. J. Appl. Math. Phys. **39**, 699–712 (1988)
62. Roberts, P.H.: Convection in a self-gravitating fluid sphere. Mathematika **12**, 128 (1965)
63. Roberts, P.H.: On the thermal instability of a rotating-fluid sphere containing heat sources. Philos. Trans. R. Soc. A **263**, 93–117 (1968)
64. Schmalzl, J., Breuer, M., Hansen, U.: On the validity of two-dimensional numerical approaches to time-dependent thermal convection. Europhys. Lett. **67**(3), 390–396 (2004)
65. Schubert, G., Glatzmaier, G.A., Travis, B.: Steady, three-dimensional, internally heated convection. Phys. Fluids A **5**(8), 1928–1932 (1993)
66. Schwiderski, E.W., Schwab, H.J.A.: Convection experiments with electrolytically heated fluid layers. J. Fluid Mech. **48**(4), 703–719 (1971)
67. Shi, B.C., Guo, Z.L.: Thermal lattice BGK simulation of turbulent natural convection due to internal heat generation. Int. J. Mod. Phys. B **17**(2), 173–177 (2003)
68. Siggia, E.D.: High Rayleigh number convection. Annu. Rev. Fluid Mech. **26**, 137–168 (1994)
69. Sotin, C., Labrosse, S.: Three-dimensional thermal convection in an iso-viscous, infinite Prandtl number fluid heated from within and from below: applications to the transfer of heat through planetary mantles. Phys. Earth Planet. Inter. **112**, 171–190 (1999)
70. Sparrow, E.M., Goldstein, R.J., Jonsson, V.K.: Thermal instability in a horizontal fluid layer: effect of boundary conditions and non-linear temperature profile. J. Fluid Mech. **18**(04), 513–528 (1964)
71. Steinberner, U., Reineke, H.H.: Turbulent buoyancy convection heat transfer with internal heat sources. In: Proceeding of 6th International Heat Transfer Conference, vol. 2, pp. 305–310 (1978)
72. Stevens, R.J.A.M., van der Poel, E.P., Grossmann, S., Lohse, D.: The unifying theory of scaling in thermal convection: the updated prefactors. J. Fluid Mech. **730**, 295–308 (2013)
73. Straughan, B.: Continuous dependence on the heat source and non-linear stability for convection with internal heat generation. Math. Methods Appl. Sci. **13**, 373–383 (1990)
74. Straus, J.M.: Penetrative convection in a layer of fluid heated from within. Astrophys. J. **209**, 179–189 (1976)
75. Takahashi, J., Tasaka, Y., Murai, Y., Takeda, Y., Yanagisawa, T.: Experimental study of cell pattern formation induced by internal heat sources in a horizontal fluid layer. Int. J. Heat Mass Transf. **53**(7-8), 1483–1490 (2010)
76. Tasaka, Y., Takeda, Y.: Effects of heat source distribution on natural convection induced by internal heating. Int. J. Heat Mass Transf. **48**(6), 1164–1174 (2005)
77. Tasaka, Y., Kudoh, Y., Takeda, Y., Yanagisawa, T.: Experimental investigation of natural convection induced by internal heat generation. J. Phys. Conf. Ser. **14**, 168–179 (2005)
78. Thirlby, R.: Convection in an internally heated layer. J. Fluid Mech. **44**(04), 673–693 (1970)
79. Tritton, D.J., Zarraga, M.N.: Convection in horizontal layers with internal heat generation. Experiments. J. Fluid Mech. **30**(01), 21–31 (1967)
80. Tveitereid, M.: Thermal convection in a horizontal fluid layer with internal heat sources. Int. J. Heat Mass Transf. **21**, 335–339 (1978)
81. Tveitereid, M., Palm, E.: Convection due to internal heat sources. J. Fluid Mech. **76**(03), 481–499 (1976)
82. van der Poel, E.P., Stevens, R.J.A.M., Lohse, D.: Comparison between two- and three-dimensional Rayleigh-Bénard convection. J. Fluid Mech. **736**, 177–194 (2013)
83. Vel'tishchev, N.F.: Convection in a horizontal fluid layer with a uniform internal heat source. Fluid Dyn. **39**(2), 189–197 (2004)
84. Wörner, M., Schmidt, M., Grötzbach, G.: Direct numerical simulation of turbulence in an internally heated convective fluid layer and implications for statistical modeling. **35**(6), 773–797 (1997)

Printed in the United States
By Bookmasters